Finite Element Methods for Engin

T0237839

J. Chaskalovic

Finite Element Methods
for Engineering Sciences

Theoretical Approach
and Problem Solving Techniques

 Springer

J. Chaskalovic
Associate Professor

Ariel University Center of Samaria   and   University Pierre and Marie Curie (Paris VI)
P.O.B 3   Institut Jean le Rond d'Alembert
40700 Ariel   4 Place Jussieu
Israel   75252 Paris Cedex 05 France

English translation and extended edition of the French book: "Méthode des éléments finis pour les sciences de l'ingénieur", Lavoisier 2004, 2-7430-0708-7.

ISBN 978-3-642-09520-7   e-ISBN 978-3-540-76343-7

DOI 10.1007/978-3-540-76343-7

© 2010 Springer-Verlag Berlin Heidelberg

Cover design: WMXDesign GmbH, Heidelberg

Printed on acid-free paper

9 8 7 6 5 4 3 2 1

springer.com

To my mother,
Who was passionate about discovering
the Universe and Science.
For all those years of sacrifice and collusion.

Preface

This book is dedicated to the use of the finite elements method for the approximation of equations having partial derivatives. It resumes part of the curriculum leading to the certificate in "Numerical Methods for Mechanics" taught by the author since the past twelve years as part of the graduate studies in Mechanics at the University of Pierre and Marie Curie (Paris VI).

Numerical Analysis has undergone a spectacular development during these past forty years. It is most probably related to the boom in Information Technology which has literally invaded the planet and provided, *de facto*, calculation capacities undreamt of up to now.

This mathematical knowledge field – Numerical Analysis – may be characterised as the "mathematic of mathematics" in the same line of thought as that of a "police of polices".

Indeed, as soon as a mathematical technology cannot be applied within the industrial applications due to operational inadequacy, numerical analysis takes over and finds the solution by identifying the best adapted approximation process.

From there on, all other mathematical branches may be used to "force a passage" and to estimate a solution by often combining shrewdness and lucidity within a stringent mathematical framework.

Numerical analysis is best and most often applied to the approximation of equations having partial derivatives as a major support to the modelling of real systems.

Whether it be applications in physics or in mechanics, in economy, marketing or in the field of finance, the phenomenological translation of the system under study often leads to the resolution of equations having partial derivatives.

This justifies the invention of numerous methods to solve such equations. The finite elements method, the finite volumes method, the singularity or integral method, the spectral methods and the variational finite differences method are some of the most popular methods.

However, the finite elements are those that definitely and drastically changed the world of numerical approximation of equations having partial derivatives. Having an exceptional flexibility, the finite elements undoubtedly constitute the approximation method that is mostly used in solving mathematical models in engineering sciences.

Considering the mathematical technicality required to apply finite elements, many authors specialised in numerical analysis, including Pierre Arnaud Raviart (8), presented the subject at higher level university teaching reserved to students who possess the mathematical prerequisites, particularly in function analysis, essential for the theoretical initiation to the finite elements method.

Other student populations who have followed a curriculum not specialised in mathematics – specially graduate and postgraduate students in Physics or in Mechanics and the Graduate Engineering Schools who are users of the mathematical tool at different degrees – may get recourse to Daniel Euvrard's (5) book which was written in the 90's and that offers a version of a course adapted to students who are unfamiliar with the tools for functional analysis.

The superiority of those two manuals reflects the quality of the teachings of Numerical Analysis by Pierre Arnaud Raviart and subsequently by Daniel Euvrard at the Training and Research Unit in Mechanics of the Pierre and Marie Curie University.

As far as this piece of work is concerned, its necessity and its core content have been greatly influenced by the strong interaction between components of the author's teaching activities in mechanics, at graduate level, consisting of Numerical Methods applied to Mechanics and to the mechanics of deformable solids at the Pierre and Marie Curie University.

Indeed, the author was motivated by the will to pursue the initiative set by the two authors mentioned above by contributing to a new balance between a selective specialist reading and one dedicated to "operational aspects while skimming over the mathematical aspects", as stated by Daniel Euvrard ([5], p.198).

The author's training and awareness on all topics dealing with numerical analysis was greatly influenced by Professor Gérard Tronel, a specially active and passionate member of the team teaching numerical analysis at the Jacques Louis Lions laboratory of the Pierre and Marie Curie University (Paris VI).

The author benefited from Professor Tronel's significant educational methods and experience as a student and, later, as a colleague and friend, in bringing about the new balance offered in this book.

The author's warm thanks are conveyed to him for his contribution.

Graduate students in Mechanics of the Pierre and Marie Curie University are the ones who have followed this novel presentation within the framework of unidimensional applications of the resistance of materials.

The present work takes up these examples again and extends them to other applications.

Having identified targeted tools for functional analysis, as exposed without any demonstration, the problems dealing with the existence, uniqueness and regularity of weak solutions and their equivalence with strong solutions have been examined through the display and use of the result of this identification.

Following this perspective, the present work is composed of a Summary of Courses on finite elements in addition to Daniel Euvrard's [5] work, and of various solutions demonstrating these techniques of functional analysis while, at the same time, tackling the construction of nodal equations characteristic of numerical implementation of the finite elements method.

Moreover, a special emphasis has been laid on the presentation of the application of assembly techniques illustrated in the problems related to the Resistance of Materials.

The author seizes this opportunity to pay tribute to the memory of Claude Kammoun who initiated him to these techniques within the framework of Resistance of Materials.

Finally, this work would never have been published without Benoît Goyeau and Cédric Croizet's proofreading of the different examination subjects at graduate level Mechanics that constitute a major part of this book. The author conveys his sincere thanks to both of them also to Dr. Arnaud Chauvière for his efficient advices in Latex Programmation.

30th September 2008   Prof. J. Chaskalovic
                                                                        Associate Professor
                                           Ariel University Center of Samaria
                               and University Pierre and Marie Curie (Paris VI)

Contents

Chapter 1

Summary of Courses on Finite Elements

1.1 Some Essential Mathematical Tools

In this section, before introducing the finite elements method and its applications, some essential tools are presented to facilitate understanding and manipulation of this particularly famous and efficient technique for the numerical analysis of equations having partial derivatives.

Keeping in mind the objectives of this book whose aim is to familiarise the reader at various degrees with a global methodology of approximation of equations having partial derivatives related to finite elements, this chapter presents the formulas for vector analysis as well as the main Hilbert and functional analysis theorems that may be applied to problems subsequently worked out in this book.

1.1.1 Adapted Functional Spaces and Their Properties

This section recalls the definition of certain functional spaces that would be used to state certain fundamental results of this chapter or to be directly used in the exercises of following chapters.

▶ **Definition 1.**

Let Ω be an open domain of $\mathbf{R}^n$ and the Sobolev space $H^1(\Omega)$ is defined as:

$$H^1(\Omega) = \left\{ v : \Omega \to \mathbf{R}, v \in L^2(\Omega), \frac{\partial v}{\partial x_i} \in L^2(\Omega), (i = 1, n) \right\}. \tag{1.1}$$

Definition 1 is then generalized by introducing the Sobolev space $H^m(\Omega)$ as follows:

▶ **Definition 2.**

$\forall m \in \mathbf{N}$ and the result is:

$$H^m(\Omega) = \left\{ v : \Omega \subset \mathbf{R}^n \rightarrow \mathbf{R}, v \in L^2(\Omega), \frac{\partial^k v}{\partial x_{i_1} \dots \partial x_{i_k}} \in L^2(\Omega), \forall k = 0, m \right\} . \tag{1.2}$$

▶ **Theorem 1.**

For any integer m, the Sobolev space $H^m(\Omega)$ is a Hilbert space.

Space $\mathscr{D}(\Omega)$, for which the notion of support (noted *Supp v*) is introduced as the smallest closed subset containing all the points where a given function v is non-zero, is another functional space essential for the functional analysis of equations having partial derivatives:

$$Supp\, v \equiv \overline{\{x \in \mathbf{R}^n / v(x) \neq 0\}}^{\mathbf{R}^n} . \tag{1.3}$$

To illustrate the notion of support, consider the example of a function of a real variable defined as:

$$H(x) = \left| \begin{array}{l} 1 \text{ given } 0 < x < 1, \\ 0 \text{ then } . \end{array} \right. \tag{1.4}$$

In this case, the function H is non-zero at the open domain $]0,1[$ but having closed interval $[0,1]$ as support:

$$Supp\, H \equiv \overline{\{x \in \mathbf{R} / H(x) \neq 0\}}^{\mathbf{R}} = [0,1] .$$

Space $\mathscr{D}(\Omega)$ is therefore defined as:

▶ **Definition 3.**

$$\mathscr{D}(\Omega) = \{ v : \Omega \subset \mathbf{R}^n \rightarrow \mathbf{R}, v \in C^\infty(\Omega), Supp\, v \subset \Omega \} . \tag{1.5}$$

Terminology: The $\mathscr{D}(\Omega)$ space is the space of functions C^∞ over Ω with a compact support *strictly* included in Ω.

The following fundamental density theorem is thus obtained:

▶ **Theorem 2.**

Space $\mathscr{D}(\Omega)$ is dense in $L^2(\Omega)$.

Finally, the closure of space $H^1(\Omega)$ in $\mathscr{D}(\Omega)$ is associated to the former and is noted as $H_0^1(\Omega)$. The following definition and property are thus obtained:

▶ **Definition 4.**

$$H_0^1(\Omega) = \overline{H^1(\Omega)}^{\mathscr{D}(\Omega)} . \tag{1.6}$$

The following result is then shown:

▶ **Theorem 3.**

$$H_0^1(\Omega) = \left\{ v : \Omega \subset \mathbf{R}^n \to \mathbf{R}, v \in H^1(\Omega), v = 0 \text{ on } \partial\Omega \right\} . \tag{1.7}$$

1.1.2 The Essential Initial Results Never to be Ever Forgotten!

The Green formula and a principal application for the variational formulations are proposed in this section.

1.1.2.1 The Green formula

▶ **Theorem 4.**

Let Ω be an open-bounded domain of $\mathbf{R}^n$ with continuous boundary $\partial\Omega = \Gamma$, only admitting discontinuities of the first kind for the tangent vector (i. e. typical angular points). Given that u and v are two functions of the defined variables $(x_1, \ldots, x_n)$ on Ω having real values and belonging to $C^1(\Omega) \cap C^0(\overline{\Omega})$. The result is:

$$\int_\Omega \frac{\partial u}{\partial x_i} \cdot v \, d\Omega = -\int_\Omega u \cdot \frac{\partial v}{\partial x_i} \, d\Omega + \int_{\partial\Omega} u \cdot v \, n_i \, d\Gamma , \tag{1.8}$$

where n_i denotes the component according to the ith coordinate x_i of normal external vector $\mathbf{n}$ to open domain Ω.

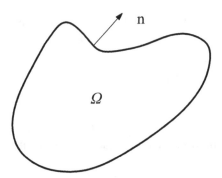

Fig. 1.1 Integration domain Ω and normal external **n**

It is to be noted that the Green formula (1.8) is nothing but a generalization of the formula of integration by parts in dimension 1.

Moreover, the application of the Green formula to functions u and v, possessing a weaker regularity than those mentioned above, is possible.

Indeed, given that u and v belong to Sobolev space $H^1(\Omega)$ then formula (1.8) is licit.

1.1.2.2 A variation of the Green formula

▶ **Theorem 5.**

Let Ω be an open-bounded domain of $\mathbf{R}^n$ with continuous boundary $\partial\Omega = \Gamma$ only admitting discontinuity of the first kind for the tangent vector (i. e.: typical angular points). Given that u and v are two functions of the defined variables $(x_1,\ldots,x_n)$ on Ω having real values such that $u \in C^2(\Omega)\cap C^1(\overline{\Omega})$ and $v \in C^1(\Omega)\cap C^0(\overline{\Omega})$, the result obtained is:

$$\int_\Omega \Delta u \cdot v \, d\Omega = -\int_\Omega \boldsymbol{\nabla} u \cdot \boldsymbol{\nabla} v \, d\Omega + \int_{\partial\Omega} \frac{\partial u}{\partial n} v \, d\Gamma , \qquad (1.9)$$

where n denotes the normal vector external to open domain Ω and $\dfrac{\partial u}{\partial n}$ the projection of the gradient vector in the direction of normal n.

It is to be noted again that the use of formula (1.9) is valid for functions having a weaker regularity, namely for $u \in H^2(\Omega)$ and $v \in H^1(\Omega)$.

1.1.3 A Set of Fundamental Inequalities

This section recalls some fundamental inequalities that emerge from the analysis and that are used intensely within the framework of functional analysis of equations having partial derivatives.

1.1.3.1 Cauchy-Schwartz Inequality

▶ **Theorem 6.**

Let u and v be two functions belonging to $L^2(\Omega)$. The result obtained is:

$$\int_{\Omega} u \cdot v \, d\Omega \leq \left[\int_{\Omega} u^2 \, d\Omega \right]^{1/2} \cdot \left[\int_{\Omega} v^2 \, d\Omega \right]^{1/2}. \tag{1.10}$$

1.1.3.2 Hölder's Inequality

▶ **Theorem 7.**

Let p and q be two conjugated real numbers that satisfy: $\dfrac{1}{p} + \dfrac{1}{q} = 1$. Let u be a function belonging to $L^p(\Omega)$ and v a function belonging to $L^q(\Omega)$. The result obtained is:

$$\int_{\Omega} u \cdot v \, d\Omega \leq \left[\int_{\Omega} u^p \, d\Omega \right]^{1/p} \cdot \left[\int_{\Omega} v^q \, d\Omega \right]^{1/q}. \tag{1.11}$$

1.1.3.3 Poincaré's Inequality

▶ **Theorem 8.**

Let Ω be an open-bounded domain of $\mathbf{R}^n$ and u a function belonging to Sobolev space $H_0^1(\Omega)$. The constant $C(\Omega)$ is such that:

$$\int_{\Omega} |u|^2 \, d\Omega \leq C(\Omega) \int_{\Omega} |\boldsymbol{\nabla} u|^2 \, d\Omega. \tag{1.12}$$

Observation: *Poincaré's inequality is also valid if function u is zero only for part of boundary $\Sigma \subset \partial\Omega$. In this case, $H_{\Sigma}^1(\Omega)$ which consists of functions belonging to $H^1(\Omega)$ and that are zero on boundary Σ replaces the $H_0^1(\Omega)$ space.*

1.1.3.4 Korn's Inequality

▶ **Theorem 9.**

Let Ω be a "sufficiently regular" open-bounded domain of $\mathbf{R}^3$. Let $\mathbf{v}$ be a field of vectors defined over Ω with components $v_i, (i \in 1,2,3)$ such that v_i belongs to $H^1(\Omega)$. It is thus stated that field $\mathbf{v}$ belongs to space $\left[H^1(\Omega) \right]^3$.

Then, the constant C > 0 is such that:

$$\int_{\Omega} [\varepsilon(\mathbf{v}) \cdot \varepsilon(\mathbf{v}) + \mathbf{v} \cdot \mathbf{v}] \, d\Omega \geq C \int_{\Omega} [\nabla \mathbf{v} \cdot \nabla \mathbf{v} + \mathbf{v} \cdot \mathbf{v}] \, d\Omega, \qquad (1.13)$$

where:

$$\varepsilon(\mathbf{v}) \cdot \varepsilon(\mathbf{v}) = \frac{1}{2} \sum_{ij} \left[\left(\frac{\partial v_i}{\partial x_i} \right)^2 + \left(\frac{\partial v_i}{\partial x_j} \frac{\partial v_j}{\partial x_i} \right) \right]. \qquad (1.14)$$

It is to be noted that $\nabla \mathbf{v}$ denotes the second order tensor of components $\dfrac{\partial v_i}{\partial x_j}$, $1 \leq i, j \leq 3$.

1.1.4 Elementary Concepts on Distributions

The Sobolev spaces $H^m(\Omega)$ were introduced in paragraph 1.1.1 according to definition 2. The elements of these spaces require an essential observation related to their intrinsic nature.

Indeed when considering the elements of space $H^1(\Omega)$, it is observed that it should relate to functions whose square as well as the square of each partial derivative can be integrated.

However, the following question needs to be asked: Should a function, whose square can be integrated, necessarily be differentiable in order to outright state that its first partial derivatives can be integrated?

The answer is obviously negative. To make things certain, consider function H introduced in (1.4).

It can be observed that it deals with a function whose square can be integrated though it cannot be differentiated at points of discontinuity when $x = 0$ and when $x = 1$.

Yet, it would be reasonable to consider this function as being an element of $H^1(\mathbf{R})$ since the H' derivative of H is a defined and always zero function, except at points of discontinuity of H where the derived function H' is not defined.

Since the two points of discontinuity, when $x = 0$ and when $x = 1$, have no bearing on the integrability of the square of derived function H' on any point of $\mathbf{R}$, it implies that H' is basically a function whose square can be integrated at any point of $\mathbf{R}$.

This is obtained by ignoring points of discontinuity of H so as to avoid problems in deriving H and assuring that the Heaviside function belongs to $H^1(\mathbf{R})$.

In other words, it seems that in this case, the digression related to the derivation of function H is not an obstacle to its belonging to space $H^1(\mathbf{R})$.

It is thus legitimate to extend these considerations by avoiding the analysis of the derivability of functions by generalising the notion of derivative to mathematical objects whose functions may prove to be a case apart.

These new mathematical objects constitute the distribution theory (1944) prodigiously invented and developed by Laurent Schwartz. "Mathematical Methods for Engineering Sciences", excellently written by Laurent Schwartz is recommended for readers seasoned in integration techniques and wishing to enrich their knowledge on this topic.

For the present work, the ambition is limited to the presentation of the basic properties of distributions by justifying the introduction of these new concepts from an analogy that may be surprising at first sight.

1st opinion: *Characterisation of a vector in a vector space of finite dimension.*

Let E be a vector space of finite dimension n and having a scalar product noted as $(.,.)_E$. Let $(e_i)_{i=1,n}$ be a basis of space E.

Any vector x belonging to E decomposed according to basis $(e_i)_{i=1,n}$ yields:

$$x = \sum_n^{i=1} x^i e_i , \qquad (1.15)$$

re-expressed according to the summation of repeated indices convention or Einstein convention in the form of:

$$x = x^i e_i . \qquad (1.16)$$

The numerical series $(x^i)_{i=1,n}$ describes the *contravariant* components of vector x in basis $(e_i)_{i=1,n}$.

The relevant dual space E^*, that is all linear forms defined over E having proven finite dimension n, is associated to vector space E.

It is noted that all linear forms belonging to E^* are necessarily continuous since dimension E is finite.

The canonical dual basis $(e^{*i})_{i=1,n}$ of E^* is then assumed to satisfy: $e^{*i}(e_j) = \delta_j^i$, (where δ_j^i is the Kronecker symbol defined as: $\delta_i^i = 1$ and $\delta_j^i = 0$ given $i \neq j$).

Thus, any linear form ω belonging to dual space E^* is expressed on the dual basis $(e^{*i})_{i=1,n}$ according to the following decomposition:

$$\omega = \omega_i e^{*i} , \qquad (1.17)$$

where the $(\omega_i)_{i=1,n}$ quantities represent the *covariant* components of the linear form ω in the dual basis $(e^{*i})_{i=1,n}$. (The Einstein convention would have been used again to simplify matters.)

For each *fixed* vector x of E, the linear form L_x defined below is then considered.

$$L_x: E \to \mathbf{R}$$
$$v \rightsquigarrow L_x(v) \equiv (x,v)_E . \qquad (1.18)$$

It is then easily established that form L_x belongs to dual space E^*. Also, L_x is expressed in basis $(e^{*i})_{i=1,n}: L_x = (L_x)_i \, e^{*i}$.

Moreover, the action of linear form L_x applied to any vector basis e_j of E is expressed as:

$$L_x(e_j) = (L_x)_i \, e^{*i}(e_j) = (L_x)_i \, \delta^i_j = (L_x)_j \equiv (x,e_j)_E . \qquad (1.19)$$

The covariant component x_j of initially fixed vector x is then defined by stating:

$$x_j \equiv (L_x)_j = (x,e_j)_E . \qquad (1.20)$$

Any vector x of E is then characterised while considering application I from E to E^* defined by:

$$I: E \to E^*$$
$$x \rightsquigarrow L_x . \qquad (1.21)$$

▶ Lemma 1.

Application I is linear and injective.

Proof:
– *Application I is linear*:

a) Let (x_1,x_2) be a pair of vectors belonging to $E \times E$.

The following is obtained:

$$I(x_1 + x_2) \equiv L_{x_1+x_2} ,$$

with:

$$L_{x_1+x_2}(v) = (x_1+x_2,v)_E = (x_1,v)_E + (x_2,v)_E = L_{x_1}(v) + L_{x_2}(v), \forall v \in E .$$

In other words:

$$I(x_1 + x_2) = I(x_1) + I(x_2) . \qquad (1.22)$$

b) Let x be any vector belonging to E and let λ be a given arbitrary real number, then:

$$I(\lambda x) \equiv L_{\lambda x} ,$$

with:

$$\forall v \in E: L_{\lambda x}(v) = (\lambda x,v)_E = \lambda(x,v)_E = \lambda L_x(v) .$$

Therefore,

$$I(\lambda x) = \lambda I(x) . \qquad (1.23)$$

– Application I is injective:
Since the application I is linear, it is sufficient to prove that its kernel is reduced to the null vector, thus showing that it is injective:

$$(\text{I is linear and injective}) \Leftrightarrow (Ker\,I = \{0\}) . \tag{1.24}$$

Hence, let x belong to $Ker\,I$. By defining the kernel of I, the following is obtained:

$$(I(x) = L_x = 0) \Leftrightarrow (\forall v \in E : L_x(v) \equiv (x,v)_E = 0) . \tag{1.25}$$

However, since the dimension of vector space E is finite, it can be deduced that only vector x, which is orthogonal for the scalar product $(.,.)_E$ to all vectors v of E, is the null vector: $x = 0$.

In other words, the kernel of application I is reduced to the null vector, implying that I is injective. *Interpretation:* The linear application I is injective, meaning that:

$$(L_x = L_y) \Rightarrow (x = y) . \tag{1.26}$$

Therefore, any known vector x belonging to E is equivalent to that of the linear form L_x associated *via* definition (1.18), that is, through the projections of vector x over all vectors v of space E.

However, the characterisation of linear form L_x *exclusively* requires the determination of its n components on the dual basis $(e^{*i})_{i=1,n}$. This is undoubtedly the result of the framework considered from the beginning of the analysis, namely, a space vector E of *finite* dimension n.

It can be noted that such a characterisation is nothing else than the projection of vector x over the n vectors which belong to the basis $(e^i)_{i=1,n}$ of E.

Conclusion: For any vector x belonging to E, the n projections of x upon the basis $(e^i)_{i=1,n}$ wholly determine this vector and correspond, *de facto*, to the known n components $(L_x)_i, (i = 1,n)$, of the linear form that is characteristic of vector x.

2nd opinion: *Extrapolation to Functions and Introduction of Distributions.*

The previous experiment concerning finite dimension suggests extrapolating the concepts of vector characterisation to certain functions which belong to the functional spaces having an *infinite* dimension.

In other words, in relation with the notations used above, it is suggested to substitute vector x of space E by a function f belonging to the functions space that bears the introduction of linear forms T_f, similar to the forms L_x, and characteristic of each vector x of E.

To propose a formalism which is broad enough, yet as simple as possible, the functions framework is studied followed by the distributions defined on $\mathbf{R}^2$.

Therefore, given that Ω indicates an open domain of $\mathbf{R}^2$, for any function f belonging to $L^2(\Omega)$, the linear form T_f is considered and is defined by:

$$T_f: \ \mathscr{D}(\Omega) \to \mathbf{R}$$
$$\varphi \rightsquigarrow T_f(\varphi) \equiv \int_\Omega f\varphi \, , \tag{1.27}$$

where $\mathscr{D}(\Omega)$ refers to the set of functions C^∞ with compact support *strictly* within Ω.

It can then be noted that definition (1.27) of the linear form T_f is licit since:

$$|T_f(\varphi)| \leq \int_\Omega |f\varphi| \leq \left(\int_\Omega f^2 \right)^{1/2} \left(\int_{Supp\varphi} \varphi^2 \right)^{1/2} , \tag{1.28}$$

where the Cauchy-Schwartz (1.10) inequality would have been used.

In so far as φ belongs to $\mathscr{D}(\Omega)$, the integral of its square is convergent over its support and the definition of T_f is obtained therefrom.

It will be observed that definition (1.27) of application T_f may be extended to the function of $L^1_{loc}(\Omega)$, (the functions space which can be completely integrated over any closed set bounded by Ω); nevertheless this functional framework requires technical adjustments that will not be considered in this particular course. Once again, the interested reader may refer to the work of Laurent Schwartz, [9].

Furthermore, the regularity C^∞ required for functions φ is significantly exaggerated, (it would have been sufficient to apply the continuity method at this stage of the presentation).

However, it will then be seen that the definition of any distribution deliberately calls for such a degree of regularity for the functions φ, upon which the effect of the distribution will be defined, but this situation has not yet been reached.

Observation and Intuitive Definition: The action of the linear form T_f over any function φ belonging to $\mathscr{D}(\Omega)$ may be interpreted as the inner product in $L^2(\Omega)$, expressed as $(.,.)_{L^2(\Omega)}$, from f by φ:

$$\int_\Omega f\varphi \equiv (f, \varphi)_{L^2(\Omega)} \, . \tag{1.29}$$

This explains why the following notation is adopted:

$$T_f(\varphi) = (f, \varphi)_{L^2(\Omega)} \equiv \langle T_f, \varphi \rangle \, . \tag{1.30}$$

The equivalent linear form L_x, which has been presented during the study of the finite dimension, is therefore available, at least in the formal form, and is characteristic of any vector x belonging to a vector space E.

To complete this analogy, it is necessary to verify the extent to which the total characterisation of any function f belonging to $L^2(\Omega)$ can be carried out by knowing the linear forms T_f.

It is therefore necessary to reconstruct the equivalent of injection I, as was defined by (1.21). This is why application J is introduced and is defined by:

$$J: \ L^2(\Omega) \rightarrow \mathscr{D}'(\Omega)$$
$$f \rightsquigarrow T_f,$$

(1.31)

where $\mathscr{D}'(\Omega)$ represents, at this stage of the construction, the set of the linear forms which are defined on $\mathscr{D}(\Omega)$.

It is then easily shown, thanks to the linear properties of the integral, that application J defined by (1.31) is itself linear.

The *injectivity* of J can now be studied.

Once again, as previously demonstrated in the case of application I defined by (1.21) and due to the linear property of J, a study of the injectivity is similar as establishing that the kernel of application J is reduced to the null element.

Let f then be a function belonging to $L^2(\Omega)$ and an element of the kernel of J. By definition, the following is obtained:

$$\left(J(f) \equiv T_f = 0\right) \Leftrightarrow \left(\int_\Omega f\varphi = 0, \forall \varphi \in \mathscr{D}(\Omega)\right).$$

(1.32)

In the present case, the difficulty to draw a conclusion lies in the fact that function f which is being sought for, belongs to space $L^2(\Omega)$ which strictly holds $\mathscr{D}(\Omega)$.

In other words, it is not, *a priori*, certain that function f which needs to be worked out, would be found among all functions φ that establish the integral equation (1.32).

Consequently, the particular case $\varphi = f$ which helps in concluding that f is equally zero cannot be directly chosen from formulation (1.32).

Moreover, and to overcome this difficulty, the result of the density theorem 2 can be applied, namely: *The space $\mathscr{D}(\Omega)$ is dense in $L^2(\Omega)$.*

This theorem is then applied in the following way: For any given function ψ belonging to $L^2(\Omega)$, there exists a sequence of functions ψ_n belonging to $\mathscr{D}(\Omega)$ such that:

$$\lim_{n\to\infty} \left[\int_\Omega |\psi_n - \psi|^2\right] = 0.$$

(1.33)

The interest of the "proximity" between the sequence of functions ψ_n and the function ψ dwells in the "contamination" of the properties of the sequence ψ_n which are passed on to the function ψ.

Indeed, it is adequate to write that the integral equation (1.32) is satisfied for the sequence of functions ψ_n belonging to $\mathscr{D}(\Omega)$:

$$\int_\Omega f\psi_n = 0, \quad \forall n \in \mathbf{N}.$$

(1.34)

Consequently, for the arbitrary function ψ belonging to $L^2(\Omega)$, the following is obtained:

$$\left| \int_\Omega f\psi \right| = \left| \int_\Omega f(\psi - \psi_n) \right| \leq \|f\|_{L^2(\Omega)} \|\psi_n - \psi\|_{L^2(\Omega)} , \qquad (1.35)$$

where the Cauchy-Schwartz (1.10) inequality has been again used.

It is sufficient to see to it that n then tends towards $+\infty$ in the inequality (1.35) so as to conclude that:

$$\int_\Omega f\psi = o, \quad \forall \psi \in L^2(\Omega) . \qquad (1.36)$$

Therefore, since f and ψ are now found in the same space $L^2(\Omega)$ within the integral equation (1.36), the particular case $\psi = f$ can assuredly be chosen, allowing as such to reach the conclusion that $f = 0$.

Consequently, the linear application J contains a kernel reduced to the null element and becomes injective:

$$\left(T_{f_1} = T_{f_2} \right) \Rightarrow (f_1 = f_2) . \qquad (1.37)$$

The characterisation of the functions f belonging to $L^2(\Omega)$ is then completed via the injection J (1.31) in the same way we introduced the adequate injection I (1.21) for any vector x belonging to a finite dimension vector space E, as previously explained.

It is now possible to define a first type of distributions which are defined on Ω.

▶ **Definition 5.**

The linear form T_f defined by (1.27) is called regular distribution associated to any function f which belongs to $L^2(\Omega)$.

Observation: To reach the definition of a distribution which is sufficiently general, it is adequate to presently keep in mind that the aim of this study is to constitute a mathematical tool that is likely to "derive" functions presenting points of discontinuity similar to the function defined by H (1.4).

In this prospect, the usual derivation properties must be preserved. Therefore, the new concept of "derivation" will have to imply the "continuity" of the distributions.

This explains why it is the proper time, at this stage, to introduce the definition of the continuous distributions.

The regular distribution support whose definition is given by (1.27) is maintained as the medium of presentation.

The continuity at the point 0 of the linear form T_f is defined:

▶ **Definition 6.**

Given φ_n a sequence of functions belonging to $\mathscr{D}(\Omega)$, it follows that the linear form T_f is continuous at the point 0 if:

$$(\varphi_n \to 0 \quad \text{in} \quad \mathscr{D}(\Omega)) \Rightarrow (T_f(\varphi_n) \to 0 \quad \text{in} \quad \mathbf{R}) . \tag{1.38}$$

Considering the linearity of the form T_f, the continuity at 0 is equivalent to any continuous point φ_0 of $\mathscr{D}(\Omega)$.

Indeed, to prove this fact, it is sufficient to express the difference $T_f(\varphi) - T_f(\varphi_0)$ in the form $T_f(\varphi - \varphi_0)$, which emphasises the reference element $\psi \equiv \varphi - \varphi_0$.

In other words, the continuity of the form T_f at the point $\varphi = \varphi_0$ is equivalent to the continuity of T_f at the point $\psi \equiv \varphi - \varphi_0 = 0$.

To ensure that the definition of the continuity 6 is complete, the convergence of a sequence of functions φ_n belonging to $\mathscr{D}(\Omega)$ needs to be specified.

▶ **Definition 7.**

A sequence of functions φ_n converge towards 0 in $\mathscr{D}(\Omega)$ given that:

1. *$\exists$ a fixed compact (independent of n) K_0 such that: $\text{Supp}\varphi_n \subset K_0$,* (1.39)

2. *$D^k \varphi_n$ uniformly converges to 0, $\quad \forall k \in N$,* (1.40)

where $D^k \varphi_n$ represents the k-differential of the sequence of functions φ_n.

The continuity which corresponds to the regular distribution T_f, as defined by (1.27) is then verified.

Hence let φ_n be a sequence of functions which belongs to $\mathscr{D}(\Omega)$ and which converges towards 0, as per the definition (1.39)–(1.40).

It is obvious that only the property of the uniform convergence of the sequence of functions φ_n is required to establish the convergence of the sequence $T_f(\varphi_n)$ in $\mathbf{R}$, according to the definition of (1.38):

$$|T_f(\varphi_n)| \leq \underset{x \in \Omega}{\text{Sup}} |\varphi_n(x)| \int_{K_0} |f| . \tag{1.41}$$

Therefore, in the case of the regular distribution T_f, the definition of the convergence in $\mathscr{D}(\Omega)$ is compatible so as to secure the continuity property which corresponds to linear forms.

At this point, it is legitimate to know why the uniform convergence of the sequence φ_n needs to be extended to the successive differentials $D^k \varphi_n$ according to the definition (1.40).

In this view, it is important to have a more global vision of the distributions, defined presently as:

▶ **Definition 8.**

A distribution T is a linear form defined on $\mathscr{D}(\Omega)$, continuous as understood in the definition 6.
Consequently the effect of the distribution T upon any function φ belonging to $\mathscr{D}(\Omega)$ can be observed, according to the following convention:

$$T(\varphi) \equiv \langle T, \varphi \rangle, \quad \forall \varphi \in \mathscr{D}(\Omega) . \tag{1.42}$$

Furthermore, all the distributions defined on Ω are referred to by $\mathscr{D}'(\Omega)$.

It can be observed that the angle bracket $\langle T, \varphi \rangle$ can no more be interpreted in the general case of a distribution T belonging to $\mathscr{D}'(\Omega)$, as the inner product in $L^2(\Omega)$ of the distribution T by the function φ.

It exclusively concerns a notation whish has been retained by analogy with the regular distributions T_f associated to the functions f belonging to $L^2(\Omega)$.

The definition 8 of a distribution T includes new mathematical tools which cannot be associated anymore to the functions f via the regular distributions T_f.

The most popular example, in this case, is the Dirac distribution δ, defined by:

$$\delta : \mathscr{D}(\mathbf{R}) \to \mathbf{R} , \quad \delta(\varphi) \equiv \langle \delta, \varphi \rangle \equiv \varphi(0) . \tag{1.43}$$

The definition of δ makes it possible to establish, by simple inspection, that it is a distribution belonging to $\mathscr{D}'(\mathbf{R})$, that is, a linear form defined on $\mathscr{D}(\mathbf{R})$ and continuous, as understood in the definition (1.38).

It can then be shown that no functions f belonging to $L^2_{\text{loc}}(\mathbf{R})$ exist such that:

$$\delta(\varphi) \equiv \varphi(0) = \int_R f\varphi, \quad \forall \varphi \in \mathscr{D}(\mathbf{R}) . \tag{1.44}$$

Indeed, proceeding by reductio ad absurdum, it is assumed that there exists a function f belonging to $L^2_{\text{loc}}(\mathbf{R})$ so that (1.44) is satisfied.

The particular case of the functions φ belonging to $\mathscr{D}(\mathbf{R})$ is chosen so that $\varphi(0) = 0$.

For each of these functions φ, there exists a function ϕ in $\mathscr{D}(\mathbf{R})$ so that:

$$\phi(x) = \frac{\varphi(x)}{x} . \tag{1.45}$$

Indeed, the only difficulty for the function ϕ dwells in its regularity in the neighbourhood of $x = 0$.

Yet, in so far as φ belongs to $\mathscr{D}(\mathbf{R})$, while having a non-zero value when $x = 0$, the expression of ϕ can be written again in the form:

$$\phi(x) = \frac{\varphi(0) + \int_0^x \varphi'(t)\,dt}{x} = \frac{\int_0^x \varphi'(t)\,dt}{x}. \tag{1.46}$$

Therefore, when x tends to 0, the following is obtained:

$$\phi(x) = \frac{\int_0^x \varphi'(t)\,dt}{x} \rightarrow \varphi'(0), \tag{1.47}$$

by applying Hôpital's rule.

However, $\varphi'(0)$ is bounded since φ is an element of $\mathscr{D}(\mathbf{R})$. Moreover, the function ϕ defined by (1.45) is bounded in the neighbourhood of $x = 0$.

In the case when $x \neq 0$, the function ϕ is C^∞ over $\mathbf{R}$, this is sufficient to ensure that it belongs to $\mathscr{D}(\mathbf{R})$.

The equation (1.44) is then expressed again by using the function ϕ defined by (1.45):

$$\varphi(0) = 0 \cdot \phi(0) = 0 = \int_R xf\phi, \quad \forall \phi \in \mathscr{D}(\mathbf{R}). \tag{1.48}$$

It is then inferred from density arguments that xf is equal to the null function, therefore implying that the function f is itself zero.

Consequently, the degenerate equality (1.44) is written as:

$$\varphi(0) = 0, \quad \forall \varphi \in \mathscr{D}(\mathbf{R}). \tag{1.49}$$

This is obviously absurd since non-zero functions when $x = 0$ exist in $\mathscr{D}(\mathbf{R})$ and thus δ does not constitute a regular distribution.

It is at present possible to define the derivation, understood as in the sense of distributions.

▶ **Definition 9.**

Let T be a distribution belonging to $\mathscr{D}'(\Omega)$. The distribution $\dfrac{\partial T}{\partial x_i}$ is defined, which is a partial derivative, understood in the sense of distributions in the direction $x_i, (i = 1, 2)$, of the distribution T, as follows:

$$\left\langle \frac{\partial T}{\partial x_i}, \varphi \right\rangle \equiv -\left\langle T, \frac{\partial \varphi}{\partial x_i} \right\rangle, \quad \forall \varphi \in \mathscr{D}(\Omega). \tag{1.50}$$

It is immediately observed that according to the definition (1.50), $\dfrac{\partial T}{\partial x_i}$ indeed constitutes a distribution belonging to $\mathscr{D}'(\Omega)$.

In fact, the properties of the distribution T confer the linear and continuity properties defined by (1.38) to $\dfrac{\partial T}{\partial x_i}$, maintaining as such its distribution status.

Observation: The definition 9 constitutes a generalisation of the form of derivation which is usual in the sense of functions.

To prove this, a function f belonging to $L^2(\Omega)$ and its associated regular distribution T_f can be considered and it can be assumed, moreover, that f is C^1 according to the *classical sense* that takes the differentiation of functions on Ω.

The partial derivative in the sense of the distributions $\dfrac{\partial T_f}{\partial x_i}$ can hence be worked out.

$\forall \varphi \in \mathscr{D}(\Omega)$, the following is obtained:

$$\left\langle \frac{\partial T_f}{\partial x_i}, \varphi \right\rangle \equiv -\left\langle T_f, \frac{\partial \varphi}{\partial x_i} \right\rangle \equiv -\int_\Omega f \frac{\partial \varphi}{\partial x_i}\, ds\,. \tag{1.51}$$

In so far as the function f has been equally accepted as C^1 on Ω in the classical sense, its regular distribution $T_{\frac{\partial f}{\partial x_i}}$ can be associated to each of its classical partial derivative $\dfrac{\partial f}{\partial x_i}$, since $\dfrac{\partial f}{\partial x_i}$ belongs to $L^2_{\mathrm{loc}}(\Omega)$.

Therefore, by using integration by parts (see Green's Formula (1.8), Chap. 1), the following is obtained:

$$\left\langle T_{\frac{\partial f}{\partial x_i}}, \varphi \right\rangle \equiv \int_\Omega \frac{\partial f}{\partial x_i} \varphi\, ds = -\int_\Omega f \frac{\partial \varphi}{\partial x_i}\, ds\,, \tag{1.52}$$

where the functions φ with a strictly included compact support in Ω have been applied.

In other words, such functions are equal to zero on the boundary $\partial\Omega$ of Ω. This explains the integral absence of a boundary in the integration by parts (1.52).

In the end, when bringing (1.51) and (1.52) closer, the following is obtained:

$$\frac{\partial T_f}{\partial x_i} = T_{\frac{\partial f}{\partial x_i}}, \quad \text{in} \quad \mathscr{D}'(\Omega)\,. \tag{1.53}$$

The interpretation of the equation (1.53) is worked out, as shown below:

The derivative $\dfrac{\partial T_f}{\partial x_i}$ is usually referred to as "the derivative of f", understood as in the sense of distributions since it is obvious that the derivation of the function f

in this sense would prove meaningless; the exclusive working out of the derivation of the distribution of T_f as a distribution makes sense.

Furthermore, the distribution $T_{\frac{\partial f}{\partial x_i}}$ is characteristic of the usual partial derivative $\frac{\partial f}{\partial x_i}$, when the injection J defined by (1.31) enables the association between this partial derivative and its regular distribution.

This becomes completely licit as soon as it is assumed that the function f is C^1, as understood in the sense of functions. In other words, its first partial derivatives are continuous over Ω and as a result, belong to $L^2_{\text{loc}}(\Omega)$.

Therefore, the equality (1.53) shows that the derivation in the sense of the distributions of a function, (i. e., its associated regular distribution T_f), coincides with the usual derivation of its functions when the distribution "is a function" which can be continuously differentiated.

The distribution of the derivative $T_{\frac{\partial f}{\partial x_i}}$ and the derivative of the distribution $\frac{\partial T_f}{\partial x_i}$ are equal.

This proves that the new derivation as well as its generalisation regarding to the classical derivative in the sense of its functions are consistent.

The example of the function H defined by (1.4) and basically belonging to $L^2(\mathbf{R})$ can again be considered to work out its derivative, as understood in the sense of distribution.

In other words, in so far as H basically belongs to $L^2_{\text{loc}}(\mathbf{R})$, it is significant to consider its regular distribution T_H defined by:

$$\forall \varphi \in \mathscr{D}(\mathbf{R}): \ \langle T_H, \varphi \rangle \equiv \int_{\mathbf{R}} H(x)\varphi(x)\,dx = \int_0^1 \varphi(x)\,dx . \tag{1.54}$$

The derivative T'_H of the regular distribution T_H is then worked out, as shown below:

$$\forall \varphi \in \mathscr{D}(\mathbf{R}): \langle T'_H, \varphi \rangle \equiv -\langle T_H, \varphi' \rangle = -\int_0^1 \varphi'(x)\,dx = \varphi(0) - \varphi(1) . \tag{1.55}$$

Hence:

$$\forall \varphi \in \mathscr{D}(\mathbf{R}): \ \langle T'_H, \varphi \rangle = \langle \delta_0 - \delta_1, \varphi \rangle , \tag{1.56}$$

where a notation (1.43) analogous to the distribution notation δ has been adopted, by specifying that the Dirac distribution is δ_0 and δ_1 characteristic of the points $x = 0$ and $x = 1$.

There consequently results:

$$\frac{dT_H}{dx} \equiv T'_H = \delta_0 - \delta_1, \quad \text{in} \quad D'(\mathbf{R}) . \tag{1.57}$$

Generalisation to the k-order derivation: The first partial derivative in the sense of the distributions (1.50) can be extended to the order k by introducing the k order partial derivative distribution, denoted $\dfrac{\partial^k T}{\partial x_1^{k_1} \partial x_2^{k_2}}$, according to the following definition:

$$\left\langle \frac{\partial^k T}{\partial x_1^{k_1} \partial x_2^{k_2}}, \varphi \right\rangle \equiv (-1)^{|k|} \left\langle T, \frac{\partial^k \varphi}{\partial x_1^{k_1} \partial x_2^{k_2}} \right\rangle, \quad \forall \varphi \in D(\Omega), \qquad (1.58)$$

where it has been observed that:

$$|k| = k_1 + k_2 \,.$$

The definition (1.58) underlines the fact that all the weight of the derivation in the sense of distributions is assumed by the functions φ belonging to $\mathscr{D}(\Omega)$. This facilitates the working out of the derivation in the sense of distributions, even if they are particularly irregular!

This principally accounts for the existence of the functional framework $\mathscr{D}(\Omega)$, requiring the regularity C^∞ of the functions φ, which define the effect of any distribution T belonging to $\mathscr{D}'(\Omega)$.

Observations: The definition of the Sobolev spaces (see definitions 1 and 2) consequently needs to be re-examined by considering the partial derivatives which occur in the definition of these spaces, like derivatives in the sense of distributions.

For example, when a "function" f belongs to space $H^1(\Omega)$, it is now clear that the first partial derivatives, in the sense of the distributions $\dfrac{\partial T_f}{\partial x_i}$ of the regular distribution T_f, are associated with the functions of $L^2(\Omega)$ as usual through the canonical injection J defined by (1.31).

Finally, to simplify writings and oral expressions, it is specified that function f and its regular distribution T_f are similar. It amounts to say that for Sobolev space $H^1(\Omega)$, its elements constitute the distributions f, (or the functions, involving a misuse of language, according to the authors) which belong to $L^2(\Omega)$ and whose partial derivatives $\dfrac{\partial f}{\partial x_i}$ are also elements of $L^2(\Omega)$, through injection J defined in (1.31).

1.2 Nature of the Finite Elements Method

1.2.1 For Mathematical Modelling

Developments in the field of numerical analysis during the 20th century gave rise to various methods that provided approximate solutions to equations having partial derivative.

Be it the finite differences method, the spectral methods, the finite volumes method or even the singularities method, it cannot be denied that the finite elements method is the most efficient one.

Undoubtedly, the other methods find their use in specific fields of application, but the finite elements have literally shattered the capacity of modifying the usually complex nature of problems with partial derivatives.

It is most certainly its tremendous adaptation to solve equations – whose inherent complexity, partly due to the domains of integration, when it comes to solving real problems emanating from industry – that caused the finite elements method to undergo such significant developments in the second half of the 20th century.

Indeed, this case is not an analytical solution whose application would cause a researcher or engineer to believe, nine times out of ten, that all classical techniques of problem solving would fail!

At this stage, it is necessary to consider the issues involved in the numerical approximation of equations having partial derivatives, irrespective of the method used.

In practice, the study of complex systems whose characteristics may widely differ from one problem to another, results in the representation of such systems by using what is commonly known as a model.

In relation with engineering sciences, many of these models consist of equations having partial derivatives.

It is therefore necessary to use an effective simulation tool, sufficiently accurate to ascertain the behaviour of a system under various conditions, instead of noting the damages which may occur in a real experiment.

However, in the first stage of the modelling, which is concerned with the choice and the mathematical translation of the fundamental mechanisms which regulate the evolution of the studied system, a non-zero rate of loss of information is inevitable.

Since the model essentially duplicates reality to some extent (it can be compared to the mirror of a bathroom where the image reflected, however much accurate and elegant, is only a two-dimensional reflection of an inevitably three-dimensional body!).

In the case of the mathematical modelling, this first stage of approximation may turn out to be disastrous.

Indeed, the analysis of the model obtained may very well result in the non-existence of a solution, in consideration of which the model must be revised, most probably improved by the addition of one or more additional mechanisms that might have been neglected during the introductory phase.

Moreover, the problem of the uniqueness of a solution needs to be carefully analysed at the level of the model, because if the latter were to generate several solutions, it would be wise to question the legitimacy of the multiplicity of these solutions in comparison with the behaviour of the real system.

Furthermore, the numerical methods which would be used later should also integrate this dimension of multiple approximate solutions.

This initial awareness on approximation, relative to the modelling of the real systems, should encourage the numerical analyst to proceed, with as much concentration and care, with the elaboration of a global methodology for approximate resolution, while imperatively respecting the necessary guidelines if final and significant results are sought for in relation to the reality of the system being modelled.

1.2.2 Structure and Functional Framework of Equations Having Partial Derivatives

Once the modelling phase is completed, there follows the choice of a problem solving method which may force the mathematical characteristic of the model. To achieve this, the mathematical model is manipulated into a form that is as far as possible conducive to a numerical approximation.

To give concrete expression to the demonstration, the two-dimensional problem of Laplace-Dirichlet which provides excellent examples for the following explanations is considered.

More specifically, let Ω be a bounded open domain of $\mathbf{R}^2$ and it is required to find function u defined from Ω to $\mathbf{R}$ and solution of:

$$(\mathbf{CP}) \begin{cases} -\Delta u = f & \text{in } \Omega, \\ u = 0 & \text{on } \partial\Omega, \end{cases} \tag{1.59}$$

where f is a given function.

At this stage, it is important to note that such a formulation is incomplete because neither the nature of the regularity of boundary $\partial\Omega$ of the integration domain Ω nor that of the second member f is specified though the regularity of solution u of continuous problem (**CP**) depends much upon it, as does the regularity of research perimeter V in which solution u can be considered.

In this way, for reasons that will be explained later, the integration domain Ω will be assumed to possess a boundary $\partial\Omega$ whose regularity is of the order of C^2. In other words, the curvature is a continuous function of the curvilinear abscissa that describes boundary $\partial\Omega$.

Moreover, assuming that the second member f belongs to $C^0(\Omega)$, it is then legitimate to consider the search for solutions of continuous problems (**CP**) as elements of $C^2(\Omega)$, thus ensuring that the Laplacian is itself continuous, (it then implies classical solutions).

In this case, the Poisson equation can be considered again, not in the form of a functional equation, but at each point M of Ω, in the form:

Find u belonging to $C^2(\Omega)$ which is the solution to:

$$(\mathbf{CP}) \begin{cases} -\Delta u(M) = f(M), & \forall M \in \Omega, \\ u(M) = 0, & \forall M \in \partial\Omega, \end{cases} \tag{1.60}$$

It is obvious that the second member f does not always exhibit regularity C^0. For instance, consider the case where f belongs to $L^2(\Omega)$. In this case, the Laplacian of solution u (which is equal to $-f$) must also be an element of $L^2(\Omega)$.

This is why it is necessary to find solution u to the continuous problem (**CP**) in Sobolev space $H^2(\Omega)$, because if this is the case, the Laplacian of u is indeed an element of $L^2(\Omega)$.

However, it must again be emphasized that the constitution of the research perimeter of solutions u results from a logical choice, because the assumed regularity of f requires the search for a solution u whose Laplacian belongs to $L^2(\Omega)$.

The Poisson equation can no more be considered, *a priori*, point-by-point as in the case of regularity C^0 for f but in the form of a functional equation. In the present case, the Poisson equation needs to be considered as an equality in $L^2(\Omega)$, that is, as a root mean square equality, or as an "energy" balance:

$$\left(\Delta u + f = 0 \quad \text{in} \quad L^2(\Omega)\right) \Leftrightarrow \left(\int_\Omega [\Delta u + f]^2 \, d\Omega = 0\right). \tag{1.61}$$

To conclude this explanation, it may be specified that the Poisson equation, considered as a functional equation in $L^2(\Omega)$ implies, nevertheless, that this equation may be studied, except for a set of zero measurements, at each point M of Ω.

For any reader who is not familiar with the concept of null set, a first encounter could consist in assuming that Poisson equation is true for each point M of Ω, except for an infinite number of countable points of Ω.

The Poisson equation would nevertheless still be studied as a global equation (1.61) written in $L^2(\Omega)$ rather than as a local equation (1.60).

1.2.3 Construction of a Variational Formulation

The essential principles constituting the finite elements method will now be studied. The basic idea prevailing in this method is to consider the unknown u no more as a scalar field which, at each point M of Ω associates a real number $u(M)$ that needs to be determined, but as an element belonging to a space of functions V in which different research trajectories would be contemplated so as to lead to the identification of the solution.

Concerning the approximation, it is no more required to determine a numerical sequence $(\tilde{u}_1, \ldots, \tilde{u}_N)$ which provides an approximation of the finite differences type for values $(u_1, \ldots, u_N)$ of solution u to continuous problem (**CP**) along points $M_j, (j = 1, N)$ that have been chosen on an adequate mesh and covering integration domain Ω.

However, it is more meaningful to elaborate a method that would lead to an approximation function $\tilde{u}$. It is obvious, *in fine*, that knowing solution u, or rather its approximation $\tilde{u}$, would facilitate the evaluation of $\tilde{u}$ at any point M of domain Ω and this evaluation would not be limited to a set of points lying on an already defined mesh, as is the case for finite differences.

A second major characteristic of the finite elements method is the transformation of a continuous problem (**CP**) into an integral formulation known as variational (**VP**).

To proceed, a function v, called test function, defined from Ω to **R** and describing the functional space V that will be elaborated later and that is not defined *a priori*.

Equation (1.59) is then multiplied by the test function v and the two members of the equation are integrated on Ω:

$$- \int_\Omega \Delta u \cdot v \, d\Omega = \int_\Omega f \cdot v \, d\Omega, \quad \forall v \in V . \tag{1.62}$$

Such a transformation is guided by the historic heritage from the finite elements method introduced as a generalisation of the Principle of Virtual Power in Continuum Mechanics (see *Cours de Mécanique des Solides*, [G. Duvaut]).

Indeed, the equation with partial derivatives (1.59) of the continuous formulation (**CP**) is none other than the written expression of the fundamental principle of statics, which presently expounds upon the equilibrium of an elastic membrane subjected to the density of transverse forces f that generate the displacement field u perpendicular to the membrane, while equation (1.62) represents an "energy" formulation in which the lay man in mechanics would still recognise that the second member of equation (1.62) is interpreted as the energy of external forces f in a displacement field v having an arbitrary status at this stage.

The left member of equation (1.62) corresponds to the energies of internal forces that are inherent to the deformation of elastic medium Ω.

Moreover, the transformation of the local writing of problem (**CP**) into a global or integral formulation (**VP**) is motivated by the need to reach a formalism that properly fits the concept of research trajectories in a functional space V.

This is precisely the case within an integral formulation, in so far as the functions do not directly reveal their numerical values at points M of Ω and only the concept of the "average value" of the functions is apparent.

The Green formula (1.9) of theorem 5 is then applied, thus enabling equation (1.62) to be written as:

$$\int_\Omega \nabla u \cdot \nabla v \, d\Omega - \int_{\partial\Omega} \frac{\partial u}{\partial n} v \, d\Gamma = \int_\Omega f \cdot v \, d\Omega, \quad \forall v \in V. \tag{1.63}$$

The present stage consists in the definition of the characteristics of space V.

A first point concerns the complete preservation of information between the writing of the formulation of the continuous problem (**CP**) and that of the variational formulation (**VP**).

As such, it is observed that the Dirichlet condition $u = 0$ along boundary $\partial\Omega$ of Ω cannot be analysed directly within the integral writing (1.63).

Considering that the future solution u of the variational problem (**VP**) must be one of the functions v of V, it is compulsory that all functions v of V satisfy the Dirichlet condition:

$$v = 0 \quad \text{on} \quad \partial\Omega. \tag{1.64}$$

This yields equation (1.63) written as:

$$\int_\Omega \nabla u \cdot \nabla v \, d\Omega = \int_\Omega f \cdot v \, d\Omega, \quad \forall v \in V. \tag{1.65}$$

The second point concerns the existence of the integrals of formulation (1.65). Indeed, it is essential to impose sufficient conditions of convergence to the integrals of equation (1.65).

In so far as it concerns the sufficient conditions of convergence, various functional contexts may constitute a favourable answer to the question.

For reasons that will be explained subsequently, the functional framework of the Sobolev spaces that provide all the desired properties is considered over and above the questions that will be dealt with at present.

The convergence of the second member of equation (1.65) is easily obtained by verification, via the Cauchy-Schwartz inequality:

$$\left| \int_\Omega f \cdot v \, d\Omega \right| \le \int_\Omega |f \cdot v| \, d\Omega \le \left[\int_\Omega |f|^2 \, d\Omega \right]^{1/2} \cdot \left[\int_\Omega |v|^2 \, d\Omega \right]^{1/2}. \tag{1.66}$$

Therefore, since f is a given function belonging to $L^2(\Omega)$, it is sufficient to consider that v is also an element of $L^2(\Omega)$, so as to ensure the convergence of the second member of equation (1.65).

In the case of the convergence of the first integral of the left member of equation (1.65), the absolute convergence of the integral is always taken into consideration and the Cauchy-Schwartz inequality is once again used:

$$\left| \int_\Omega \nabla u \cdot \nabla v \, d\Omega \right| \le \int_\Omega |\nabla u \cdot \nabla v| \, d\Omega \le \left[\int_\Omega |\nabla u|^2 \, d\Omega \right]^{1/2} \cdot \left[\int_\Omega |\nabla v|^2 \, d\Omega \right]^{1/2}. \tag{1.67}$$

Convergence of the first member of (1.65) is hence assured if the gradients of test function v belonging to V are compulsorily elements that belong to $L^2(\Omega)$.

In conclusion, it has been proved that the <u>sufficient</u> conditions for the convergence of the integrals of equation (1.65) are:

$$v \in L^2(\Omega) \quad \text{and} \quad \nabla v \in [L^2(\Omega)]^2 .$$

These reasons consequently explain the choice of the variational space V as the Sobolev space $H^1(\Omega)$ earlier presented in the paragraph 1.1.1 and to which the homogenous Dirichlet condition (1.64) must necessarily be added.

In other words, the following is stated:

$$V \equiv H_0^1(\Omega) \equiv \{v : \Omega \to R, v \in L^2(\Omega), \nabla v \in [L^2(\Omega)]^2, v = 0 \text{ on } \partial\Omega\} \qquad (1.68)$$

All the results when grouped enable the expression of the variational formulation (**VP**) that will be considered in the sequel:

$$(\textbf{VP}) \begin{cases} \text{Find } u \in H_0^1(\Omega) \text{ solution of:} \\ \displaystyle\int_\Omega \nabla u \cdot \nabla v \, d\Omega = \int_\Omega f \cdot v \, d\Omega, \quad \forall v \in H_0^1(\Omega) . \end{cases} \qquad (1.69)$$

1.2.4 Existence, Uniqueness and Regularity of a Weak Solution

Obtaining existence and uniqueness results for solutions to differential equations or to partial differential equations or to variational equations is a totally open topic. The complexity of such results depends on the nature and structure of the equation, or system of equations, under study.

Concerning variational formulations, there is a sufficient general formalism for which, under some conditions, the existence and uniqueness of the solution may be guaranteed.

It is the object of the Lax-Milgram theorem that is pointed out in the following form:

▶ **Theorem 10.**

Let V be a Hilbert space in relation to a given norm $\|.\|$, $a(.,.)$ a bilinear form defined on $V \times V$ and L a linear form defined on V verifying the following properties:

1. $a(.,.)$ is continuous: $\exists C_1 > 0$ such that: $|a(u,v)| \le C_1\|u\| \cdot \|v\|, \forall(u,v) \in V \times V$.

2. $a(.,.)$ is V-elliptical: $\exists C_2 > 0$ such that: $a(v,v) \ge C_2\|v\|^2, \forall v \in V$.

3. L is continuous: $\exists C_3 > 0$ such that: $|L(v)| \le C_3\|v\|, \forall v \in V$.

Then, there is one and only one solution u belonging to V, solution to the variational problem:

$$\text{Find } u \in V \text{ solution of: } a(u,v) = L(v), \quad \forall v \in V . \qquad (1.70)$$

Observations:

i) The 3 constants $C_i, (i = 1, 2, 3)$ intervening in each of the three clauses of the Lax-Milgram theorem must absolutely be *independent* from the generic element v covering the space V.

ii) It is essential to note that during the application of the Lax-Milgram theorem, all the properties required necessitate the use of a unique norm of the Hilbert space V, (noted as $\|.\|$), mainly in order to establish its Hilbertian property.

Yet, it is possible that for the sake of convenience, a change of norms is required to prove one of the properties of the Lax-Milgram theorem.

In this case, it is appropriate to ensure that all the applied norms are equivalent, namely:

Given $\|.\|_1$ and $\|.\|_2$ two appropriate norms for space V, it should be established that there are two constants α and β that are strictly positive and *independent* from v such that:

$$\forall v \in V : \alpha\|v\|_2 \leq \|v\|_1 \leq \beta\|v\|_2 .$$

The following two lemmas are fundamental for the *a priori* analysis of the regularity of weak solutions to a variational formulation having a one space dimension. Their demonstrations may be consulted in the work of H. Brézis [1].

▶ Lemma 2.

Let I be an open interval of $\mathbf{R}$ *and f a function belonging to* $L^1_{loc}(I)$ *verifying:*

$$\int_I f(x)\varphi(x)\,dx = 0, \quad \forall \varphi \in C^1_0(I) . \tag{1.71}$$

then: $f = C^{te}$ almost everywhere.

$C^1_0(I)$ refers to the defined functions and C^1 to those defined over the interval I, having a compact support and strictly included in I.

▶ Lemma 3.

Consider $g \in L^1_{loc}(I)$; *for* y_0 *fixed in I, the following is expressed:*

$$v(x) = \int_{y_0}^x g(t)\,dt, \quad \forall x \in I . \tag{1.72}$$

Then $v \in C(I)$ *(given that I is a bounded interval then v belongs to* $H^1(I)$*) and*

$$\int_I v\varphi' = -\int_I g\varphi, \quad \forall \varphi \in C^1_0(I) . \tag{1.73}$$

Finally, a trace theorem that is very useful for the application of the Lax-Milgram theorem is recalled, mainly in the framework of the Laplacian-Neumann-Dirichlet problem.

▶ **Theorem 11.**

Assume that Ω is an open bounded domain of $\mathbf{R}^2$, having boundary $\Gamma = \partial\Omega$, which is "sufficiently regular", (at least C^1-per piece).

Application γ defined by:

$$\gamma\colon H^1\,(\Omega) \rightarrow L^2(\Gamma) \tag{1.74}$$
$$v \rightsquigarrow v|_\Gamma ,$$

is linear continuous.

In other words, there is a constant $C > 0$ independent from v, such that:

$$\forall v \in H^1(\Omega)\colon \|v\|_{L^2(\Gamma)} \leq C\|v\|_{H^1(\Omega)} . \tag{1.75}$$

Application to the Laplacian-Dirichlet Problem.

A first application of the Lax-Milgram theorem is proposed in order to establish the existence and uniqueness of the solution to the variational formulation (**VP**) defined by (1.69), associated with the continuous problem (**CP**) defined by (1.59) when the given data f belongs to $L^2(\Omega)$.

Application of the Lax-Milgram theorem requires the identification of space V, the bilinear form $a(.,.)$ and that of the linear form $L(.)$.

Variational formulation (**VP**) defined by (1.69) suggests the introduction of the following quantities:

Let V be the search space of solution u to the variational problem defined by: $V = H_0^1(\Omega)$.

Space $H_0^1(\Omega)$ is provided with the natural norm $\|.\|_{H^1(\Omega)}$ of functions belonging to $H^1(\Omega)$.

Thus, $\forall v \in H^1(\Omega)$, the following is written:

$$\|v\|_{H^1(\Omega)}^2 \equiv \int_\Omega v^2\,d\Omega + \int_\Omega \left(\frac{\partial v}{\partial x}\right)^2 d\Omega + \int_\Omega \left(\frac{\partial v}{\partial y}\right)^2 d\Omega . \tag{1.76}$$

This norm is Hilbertian for space $H^1(\Omega)$, (see. [8]), as well as for $H_0^1(\Omega)$ as a closed vectorial subspace in $H^1(\Omega)$.

Let a be the bilinear form defined by:

$$a: \; V \times V \rightarrow R$$
$$(u,v) \rightsquigarrow a(u,v) \equiv \int_{\Omega} \nabla u \cdot \nabla v \, d\Omega \,. \tag{1.77}$$

Likewise, let L be the linear form defined by:

$$L: \; V \rightarrow R$$
$$v \rightsquigarrow L(v) \equiv \int_{\Omega} f v \, d\Omega \tag{1.78}$$

Thus, variational formulation (**VP**) defined by (1.69) is written in the form:

$$\text{Find } u \in V \text{ solution of}: a(u,v) = L(v), \quad \forall v \in H_0^1(\Omega) \,. \tag{1.79}$$

Then, a verification of the clauses of the Lax-Milgram theorem 10 is carried out.

1. $a(.,.)$ is a continuous bilinear form:

The bilinearity of form $a(.,.)$ is obvious.

As for its continuity, consider any two elements u and v belonging to $H_0^1(\Omega)$.

The following is obtained:

$$|a(u,v)| \leq \int_{\Omega} |\nabla u \cdot \nabla v| \leq \left(\int_{\Omega} |\nabla u|^2 \right)^{1/2} \cdot \left(\int_{\Omega} |\nabla v|^2 \right)^{1/2}, \tag{1.80}$$

where the Cauchy-Schwartz inequality would have been used.

However,

$$\int_{\Omega} |\nabla u|^2 = \int_{\Omega} \left[\left(\frac{\partial u}{\partial x} \right)^2 + \left(\frac{\partial u}{\partial y} \right)^2 \right] = \left\| \frac{\partial u}{\partial x} \right\|_{L^2(\Omega)}^2 + \left\| \frac{\partial u}{\partial y} \right\|_{L^2(\Omega)}^2, \tag{1.81}$$

where $\|.\|_{L^2(\Omega)}$ refers to the natural norm in $L^2(\Omega)$, namely:

$$\forall u \in L^2(\Omega): \|u\|_{L^2(\Omega)} \equiv \left(\int_{\Omega} |u|^2 \right)^{1/2}. \tag{1.82}$$

The following is then inferred:

$$\int_{\Omega} |\nabla u|^2 \leq \|u\|_{H^1(\Omega)}^2 \,. \tag{1.83}$$

Inequality (1.80) then leads to:

$$|a(u,v)| \leq \|u\|_{H^1(\Omega)} \cdot \|v\|_{H^1(\Omega)} , \tag{1.84}$$

and the continuity constant C_1 of theorem 10 is basically equal to one.

2. $a(.,.)$ is a V-elliptical form:

In order to establish the V-ellipticity of the bilinear $a(.,.)$ form, the quantity $a(v,v)$ defined from (1.77) needs to be minorated.

Also, any function $v \in H_0^1(\Omega)$ yields:

$$a(v,v) = \int_\Omega |\nabla v|^2 = \left\|\frac{\partial v}{\partial x}\right\|_{L^2(\Omega)}^2 + \left\|\frac{\partial v}{\partial y}\right\|_{L^2(\Omega)}^2 . \tag{1.85}$$

In order to obtain a lower bound of $a(v,v)$ in relation to the $H^1(\Omega)$ norm, it is pointed out that for all functions v belonging to $H_0^1(\Omega)$, the Poincaré inequality (1.12) is available.

In other words, a constant $C(\Omega) > 0$ exists such that:

$$\int_\Omega |v|^2 \, d\Omega \leq C(\Omega) \int_\Omega \left|\nabla v\right|^2 d\Omega . \tag{1.86}$$

To each side of inequality (1.86), the square of norm $L^2(\Omega)$ of the module of ∇v is added so as to yield the square of norm $H^1(\Omega)$ of function v:

$$\|v\|_{H^1(\Omega)}^2 \equiv \|v\|_{L^2(\Omega)}^2 + \left\|\frac{\partial v}{\partial x}\right\|_{L^2(\Omega)}^2 + \left\|\frac{\partial v}{\partial y}\right\|_{L^2(\Omega)}^2 \tag{1.87}$$

$$\leq (1 + C(\Omega)) \left[\left\|\frac{\partial v}{\partial x}\right\|_{L^2(\Omega)}^2 + \left\|\frac{\partial v}{\partial y}\right\|_{L^2(\Omega)}^2\right] \tag{1.88}$$

$$\leq (1 + C(\Omega)) a(v,v) . \tag{1.89}$$

It then becomes:

$$a(v,v) \geq C_2 \|v\|_{H^1(\Omega)}^2 , \tag{1.90}$$

where the V-ellipticity constant C_2 is defined by: $C_2 = \dfrac{1}{1 + C(\Omega)}$.

3. $L(.)$ is a continuous linear form:

Once again, the linearity of form L is obvious.

Control of linear form L is quite simple being given that f is a function belonging to $L^2(\Omega)$:

$$|L(v)| \leq \int_\Omega |fv| \, d\Omega \leq \|f\|_{L^2(\Omega)} \cdot \|v\|_{L^2(\Omega)} \leq \|f\|_{L^2(\Omega)} \cdot \|v\|_{H^1(\Omega)} . \tag{1.91}$$

Continuity constant C_3 of linear form L is thus equal to $\|f\|_{L^2(\Omega)}$.

Result: According to the Lax-Milgram theorem, only one function belongs to $H_0^1(\Omega)$ solution of the variational formulation (**VP**) defined by (1.79).

Observations:

(i) When the given data f shows less regularity than that considered (i. e. $L^2(\Omega)$), then the tools necessary for the functional analysis of the variational formulation do not fall within the framework of this course.

The reader who is interested in this aspect of the problem may consult more specialised works such as that of H. Brezis [1] or the Robert Dautray and Jacques-Louis Lions collection [3].

(ii) In case the continuous problem (**CP**) defined by (1.59) is replaced by the Laplace-Neumann-Dirichlet problem then the boundary Γ of Ω is constituted of two complementary parts Γ_1 and Γ_2, respectively dedicated to the definition of the Dirichlet and the Neumann conditions.

In such a case, the continuous problem (**CP**) takes the following form:

$$(\mathbf{CP})\begin{cases} -\Delta u = f & \text{in} \quad \Omega\,, \\ u = 0 & \text{on} \quad \Gamma_1\,, \\ \dfrac{\partial u}{\partial n} = g & \text{on} \quad \Gamma_2\,, \end{cases} \tag{1.92}$$

where it is assumed that f and g are two given functions respectively belonging to $L^2(\Omega)$ and to $L^2(\Gamma_2)$.

As a consequence, it is easily established that the new associated variational formulation is written as:

$$(\mathbf{VP})\begin{cases} \text{Find } u \in H_{\Gamma_1}^1(\Omega) \text{ solution to:} \\ \displaystyle\int_\Omega \boldsymbol{\nabla} u \cdot \boldsymbol{\nabla} v \, d\Omega = \int_\Omega f \cdot v \, d\Omega + \int_{\Gamma_2} g \cdot v \, d\Gamma\,, \quad \forall v \in H_{\Gamma_1}^1(\Omega)\,, \end{cases} \tag{1.93}$$

where Sobolev $H_{\Gamma_1}^1(\Omega)$ space is defined by:

$$H_{\Gamma_1}^1(\Omega) \equiv \left\{ v\colon \Omega \to \mathbf{R}, v \in L^2(\Omega), \boldsymbol{\nabla} v \in \left[L^2(\Omega)\right]^2, v = 0 \text{ on } \Gamma_1 \right\}. \tag{1.94}$$

Observations:

(i) When the Dirichlet given data on Γ_1 is not homogeneous, the plot technique [3], which may prove to be very technical, enables the transformation of the non-homogeneous problem into a homogeneous formulation identical to the one presented by (1.92).

(ii) The Lax-Milgram theorem applied to variational problem (1.93) is performed in an analogous manner to the one presented for variational formulation (1.69), that is, associated to the Laplace-Dirichlet problem.

However, some substantial modifications need to be performed to obtain continuity in the linear form $L(.)$.

In fact, in this case, the action of form L on any function v belonging to $H^1_{\Gamma_1}(\Omega)$ is expressed as:

$$L(v) \equiv \int_{\Omega} f \cdot v \, d\Omega + \int_{\Gamma_2} g \cdot v \, d\Gamma, \quad \forall v \in H^1_{\Gamma_1}(\Omega) . \tag{1.95}$$

The control of $L(v)$ is then carried out using:

$$|L(v)| \leq \int_{\Omega} |f \cdot v| \, d\Omega + \int_{\Gamma_2} |g \cdot v| \, d\Gamma , \tag{1.96}$$

$$\leq \|f\|_{L^2(\Omega)} \|v\|_{L^2(\Omega)} + \|g\|_{L^2(\Gamma_2)} \|v\|_{L^2(\Gamma_2)} . \tag{1.97}$$

Thus, a new difficulty results from the application of the g Neumann condition defined on the Γ_2 boundary.

Since, control of $L(v)$ should be performed only in relation to norm $H^1(\Omega)$ of function v. This is why the term resulting from the Neumann condition and providing a measure of v for norm $L^2(\Gamma_2)$ should consequently be modified.

The trace theorem 11 mentioned above is the one that would enable a control over $L(v)$ in relation to the only measure of function v for norm $H^1(\Omega)$.

It is to be noted that C_4 is the continuity constant of the trace application γ defined by (1.74).

Then, inequality (1.97) may be modified as follows:

$$|L(v)| \leq \|f\|_{L^2(\Omega)} \|v\|_{L^2(\Omega)} + C_4 \|g\|_{L^2(\Gamma_2)} \|v\|_{H^1(\Omega)} , \tag{1.98}$$

$$\leq C_5 \|v\|_{H^1(\Omega)}, \quad \forall v \in H^1_{\Gamma_1}(\Omega) , \tag{1.99}$$

where it would have been set that: $C_5 = \|f\|_{L^2(\Omega)} + C_4 \|g\|_{L^2(\Gamma_2)} .$

These are the essential points that needed to be specified for extending the Laplace-Dirichlet problem to that of Laplace-Neumann-Dirichlet.

Other minor modifications, that do not represent any major difficulties, concern the adaptation of the results while shifting the functional framework of $H^1_0(\Omega)$ to that of $H^1_{\Gamma_1}(\Omega)$.

This is why, once the point about the control of linear form $L(.)$ defined by (1.95) is made, the application of Lax-Milgram theorem guarantees the existence and uniqueness of solution $u \in H^1_{\Gamma_1}(\Omega)$ to the variational problem (**VP**) defined by (1.93).

1.2.5 Equivalence Between Strong and Weak Formulations

An additional point concerning the equivalence between different formulations needs to be mentioned within the whole transformation process that has been presented above.

More precisely, it is not obvious to declare that any solution to variational problem (**VP**) (1.69) is a solution to continuous problem (**CP**) (1.59). Moreover, in many cases this is absolutely wrong!

The subtleties of the concept of equivalence between the two formulations may be tested by continuing to assume that the second member f is a function belonging to $L^2(\Omega)$ and it is then only necessary to believe that if the solution to the continuous problem is searched for in the Sobolev space $H^2(\Omega)$, then that of the variational problem (**VP**) is searched for in $H^1(\Omega)$ and $H^2(\Omega) \subset H^1(\Omega)$.

In other words, any solution to the continuous problem may be a solution to a variational problem with regards to its regularity, whereas, *a priori*, there is no justification for a solution to a variational problem (**VP**) to be the solution to a continuous problem (**CP**).

In fact, the concept of equivalence between the two formulations is completely dependent on the functional frameworks governing the respective areas of research for solutions to a continuous problem (**CP**) on one hand and to a variational problem (**VP**) on the other hand.

Now, the manner to establish equivalence between a variational problem (**VP**) and a continuous problem (**CP**) will be demonstrated within the framework of the Laplace-Dirichlet problem.

In an obvious manner and through construction, any solution to a continuous problem (**CP**) belonging to $H^2(\Omega)$ is a solution to the variational problem (**VP**).

Given that, *a priori* (i. e. independently of the fact that solutions are known or not), regularity properties of a solution u to the variational problem (**VP**) depend on the regularity of the second member f as well as on the geometrical properties of the boundary $\partial\Omega$ of the integration domain Ω, a "partial reciprocal" will be stated as follows:

Is a solution to the variational problem (**VP**) showing regularity properties of a continuous problem (**CP**), i. e. belonging to $H^2(\Omega)$ (and not only the regularity of $H^1(\Omega)$), the solution to the continuous problem (**CP**)?

Thus, consider u belonging to $H^2(\Omega)$ being a solution to the variational problem (**VP**). By using the Green formula, the following is obtained:

$$-\int_\Omega \Delta u \cdot v \, d\Omega + \int_{\partial\Omega} \frac{\partial u}{\partial n} v \, d\Gamma = \int_\Omega f \cdot v \, d\Omega, \quad \forall v \in H_0^1(\Omega). \qquad (1.100)$$

As, in addition, v belonging to $H_0^1(\Omega)$, the boundary integral over $\partial\Omega$ in equation (1.100) is zero.

It then becomes:

$$\int_\Omega [\Delta u + f] \cdot v \, d\Omega = 0, \quad \forall v \in H_0^1(\Omega) . \tag{1.101}$$

The necessity of introducing regularity $H^2(\Omega)$ for a solution u upon application of the Green formula would be noted and this is performed in order to guarantee the convergence of the integral that causes the Laplacien of u to intervene.

In fact, by always using the Cauchy-Schwartz inequality, an increase in control is obtained as follows:

$$\left| \int_\Omega \Delta u \cdot v \, d\Omega \right| \leq \int_\Omega |\Delta u \cdot v| \, d\Omega \leq \left[\int_\Omega |\Delta u|^2 \, d\Omega \right]^{1/2} \cdot \left[\int_\Omega |v|^2 \, d\Omega \right]^{1/2} . \tag{1.102}$$

Now the problem cropping up from equation (1.101) is considered. In fact, it would be desirable to affirm that such a family of equalities, since there are as many equations as functions v in $H_0^1(\Omega)$, should lead to:

$$\Delta u + f = 0 \quad \text{in} \quad \Omega . \tag{1.103}$$

However, the shift from the integral equation (1.101) to the equation with partial derivatives (1.103) might be justified subject to the possibility of choosing the specific function $v^* = \Delta u + f$ from all functions v of $H_0^1(\Omega)$.

In such a case, the integral of equation (1.101) for such a specific choice would be expressed as:

$$\int_\Omega |\Delta u + f|^2 \, d\Omega = 0 , \tag{1.104}$$

which would necessitate the development of the Poisson equation (1.103) (the integral of a positive or zero function can be zero only if its integrand is equally zero.)

However, function v of $H_0^1(\Omega)$ cannot directly be designated as the specific function $v^* = \Delta u + f$ since the latter does not belong to $H_0^1(\Omega)$ but belongs to $L^2(\Omega)$ only.

A density technique is applied to overcome this obstacle, i. e. a technique of "contamination" by proximity.

The key to achieve this is a density theorem that would, through neighbourhood proximity, enable the extension of the desired property (integral equation (1.101) in this case) to suitable functions of $L^2(\Omega)$.

More specifically, a density theorem would be applied to show that equation (1.101) may be written for any v belonging to $H_0^1(\Omega)$ and also for any function v belonging to $L^2(\Omega)$ in order to facilitate the choice of the specific function v that is equal to $\Delta u + f$.

Caution! This result is not trivial since the inclusion of functional spaces is not such as would enable the application of a reasoning of the type: "that which is capable of more, is capable of less"...

In fact, given that $H_0^1(\Omega)$ is included in $L^2(\Omega)$, and not the other way round, it cannot be claimed that equation (1.101) may be written, particularly for any v belonging to $L^2(\Omega)$.

Then, the density theorem 2 is applied and this enables the assertion that for any function w belonging to $L^2(\Omega)$, there exists a sequence of functions w_n of $C_0^\infty(\Omega)$ that converge towards function w along the same sense as for norm $L^2(\Omega)$.

$$\lim_{n\to\infty}\left[\int_\Omega |w_n - w|^2 \, d\Omega\right] = 0. \tag{1.105}$$

In addition, the following Sobolev embedding holds: $C_0^\infty(\Omega) \subset H_0^1(\Omega)$. It is thus legitimate to write equation (1.101) for each function of the sequence w_n as these particularly belong to $H_0^1(\Omega)$.

$$\int_\Omega [\Delta u + f] \cdot w_n \, d\Omega = 0, \quad \forall n \in \mathbf{N}. \tag{1.106}$$

It is then possible to establish the same property for functions w of $L^2(\Omega)$:

$$\left|\int_\Omega [\Delta u + f] \cdot w \, d\Omega\right| = \left|\int_\Omega [\Delta u + f] \cdot (w - w_n) \, d\Omega\right| \tag{1.107}$$

$$\leq \left[\int_\Omega |\Delta u + f|^2 \, d\Omega\right]^{1/2} \cdot \left[\int_\Omega |w_n - w|^2 \, d\Omega\right]^{1/2}. \tag{1.108}$$

It then suffices to tend n towards $+\infty$ in integrality (1.108) and to the property of convergence (1.105) so that for any function w belonging to $L^2(\Omega)$, the following is obtained:

$$\int_\Omega [\Delta u + f] \cdot w \, d\Omega = 0, \quad \forall w \in L^2(\Omega). \tag{1.109}$$

The conclusion is then immediate as mentioned before and the Poisson equation is thus satisfied in $L^2(\Omega)$ for any solution u of the variational problem (**VP**) that possesses the additional regularity of belonging to $H^2(\Omega)$.

It will again be observed that this last property is useful because Δu, being a function of $L^2(\Omega)$, enables the use of density theorem 2.

In fact, the hypothesis that consists in considering any solution to the variational problem (**VP**) in $H^2(\Omega)$ may be rejected if $\partial\Omega$ is C^2 and this would be possible if functional analysis tools pertaining to the theory of *a priori* estimation were available.

In fact, it could be demonstrated that given the second member f belongs to $L^2(\Omega)$, then any solution u to the variational problem (**VP**) belonging to $H_0^1(\Omega)$ is also an element of $H^2(\Omega)$ for Ω domains of integration whose geometry of boundary $\partial\Omega$ exhibits a C^2 regularity (see H. Brézis, [1]).

This is exactly what was assumed in the paragraph entitled "Formalism and Functional Framework of Equations Having Partial Derivatives".

By contrast, in the case of a geometry that exhibits prominent angles (see Fig. 1.2), it can be shown that the solution to the variational problem (**VP**) does not belong to $H^2(\Omega)$.

Worse, for certain geometries, there exist infinite solutions to the continuous problem (**CP**) (see M. Moussaoui, [6]) when the variational problem (**VP**) possesses one and only one solution belonging to $H_0^1(\Omega)$.

Thus, for situations exhibiting geometrical peculiarities at the $\partial\Omega$ boundary and with a view to preserve equivalence between the two formulations, it is necessary to restrict the solutions research perimeter of the continuous problem (**CP**) to space $H_0^1(\Omega)$.

It may then be shown that there exists one and only one function belonging to $H_0^1(\Omega)$ being the simultaneous solution to continuous problems (**CP**) and to variational problems (**VP**).

In other words, since f belongs to $L^2(\Omega)$, the Poisson equation would be satisfied in $L^2(\Omega)$ and thus almost everywhere.

As a consequence, solution u would have a Laplacian in $L^2(\Omega)$, however this does not imply that u belongs to $H^2(\Omega)$.

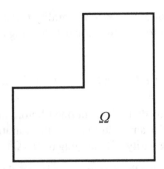

Fig. 1.2 Integration Domain Ω Exhibiting a Prominent Angle

Finally, to complete this "partial reciprocal" it should be noted that the homogeneous Dirichlet condition over $\partial\Omega$ is automatically satisfied for each solution u to the variational problem (**VP**) and being an element of space $H_0^1(\Omega)$, these functions possess an identical zero trace over $\partial\Omega$.

The aim of the demonstration was to create an awareness, may be stronger than usual, among those who, for various reasons, do not pay attention to all those questions that might appear as taken for granted when in fact they represent the surety, the credibility and the perpetuity of mathematical models applied to engineering sciences.

This process should not be misunderstood and it is only by controlling the aspects of methodological coherence that the numerical analyst may reach a position to produce a real scientific system of modelling that can be an aid to decision taking and to forecasting of the behaviour of real systems.

1.2.6 Methodology and Approximation Cascades

It is now important to study the structure of the formulation of the variational problem (**VP**) (1.69), to compare it with that of the continuous problem (**CP**) (1.59), so as to understand and identify the mechanisms that would not allow the display of methods of solving problems through analytical pathways.

A numerical approach would be successful under such conditions since all or part of structural constraints that limit the resolution of a problem would have been eliminated from whatever conceivable mathematical formulation.

A first observation is made on the continuous problem (**CP**). Two related mechanisms lead to the impossibility of resolving a problem through the analytical pathway.

The first one, immediate and obvious, resides in the complexity of combined differentiation operations on the unknown function u.

In fact, the combination of two second order partial derivatives unquestionably constitutes a barrier to the resolution and this is true for whatever expression and complexity of the second member f.

Of course, the reader who masters the techniques of solving equations with partial derivatives (see D. Euvrard, [5]) would wish to apply one or several miraculous transformations (Laplace, Fourrier,...).

However, the above neglects the fact that the second factor intervenes in the problematic of resolution and it relates to the more or less complex form of the integration domain Ω.

In fact, when Ω has a regular form (square, circular, elliptical, ...), the essential difficulty to resolve is masked. Since, in a general manner, there is necessity to read the formulation of the continuous problem (**CP**) in a different way and may be in an unusual way.

This reading requires that the continuous problem be considered as a system of non-algebraic equations made up of an infinite number of equations for an infinite number of unknowns.

Since, in fact, the Poisson (1.59) equation with partial derivatives is finally written for each point M of the domain of integration Ω.

And, as commonly known, the fact that there is an infinite number of points that constitutes the interior of Ω leads to the consideration, in general silently, of an infinite number of equations for an infinite number of unknowns that are nothing else than the values of function u at each point M of Ω!

These reasons drive the numerical analyst to transform the continuous problem into a formulation that is expressed in a finite dimension since human beings are not structured nor "equipped" to understand the infinite dimension...

The finite differences method demonstrated in various works (see D.Euvrard, [5]) performs this transformation through restriction, by introducing a mesh so as to consider only a number of finite points M_i of a grid and approximations $\tilde{u}_i$ at these points, to yield solutions to a system of algebraic equations obtained by approximation of partial derivatives after having basically used the formula of Taylor.

But then, is an acceptable alternative obtained by the transformation of the continuous problem (**CP**) into a variational problem (**VP**) as demonstrated?

In the affirmative, what would then be the advantage of this new formulation that, at first sight, seems to render the original continuous problem (**CP**) complex.

At this stage, it is to be noted that the inherent difficulty of the continuous problem (**CP**) has completely been transferred into variational formulation (**VP**).

In fact, as demonstrated above, the infinite number of unknowns and of equations associated with the Poisson equation (1.59) determine one of the main obstacles to resolution.

In the present case of variational formulation (1.69), this concept of infinite dimension is always present and it concerns the research space V which is here represented by $H_0^1(\Omega)$ and, as a consequence, the infinite number of equations present in variational formulation (1.69).

This is why the method of approximation of finite elements, also known as the method of Galerkin, consists in considering a sub-space $\tilde{V}$ having a finite dimension. State K_h to be the dimension of space K_h.

The shift from variational problem (**VP**) to the approximate variational problem $\widetilde{(\textbf{VP})}$ is performed by substituting the pair of functions (u, v) belonging to $V \times V$ by their approximations $(\tilde{u}, \tilde{v})$ belonging to $\tilde{V} \times \tilde{V}$.

Thus $\widetilde{(\textbf{VP})}$ is written as:

$$\widetilde{(\textbf{VP})} \begin{cases} \text{Find } \tilde{u} \in \tilde{V} \text{ solution of:} \\ \displaystyle\int_\Omega \boldsymbol{\nabla}\tilde{u} \cdot \boldsymbol{\nabla}\tilde{v}\,\mathrm{d}\Omega = \int_\Omega f \cdot \tilde{v}\,\mathrm{d}\Omega, \quad \forall \tilde{v} \in \tilde{V}\,. \end{cases} \tag{1.110}$$

Care should be taken to avoid the misleading simplicity of the approximation process since the approximate variational formulation $\widetilde{(\mathbf{VP})}$ is not simply a writing composition in relation to formulation $(\mathbf{VP})$.

On the contrary, it is a real progress in the capacity to resolve variational problem $(\mathbf{VP})$ by approximation but it also relates to a loss of information that should be estimated subsequently.

In order to really appreciate the critical progress that this represents in terms of resolution, we introduce a basis $(\varphi_i)_{i=1,K_h}$ of the approximation space $\tilde{V}$ we consider which a finite dimensional vector space.

In this case, unknown $\tilde{u}$ may be broken down on the basis of functions φ_i as below:

$$\tilde{u} = \sum_{j=1}^{K_h} \tilde{u}_j \varphi_j . \tag{1.111}$$

In other words, since equation (1.110) is true, $\forall \tilde{v} \in \tilde{V}$, each basis functions φ_i, $(i = 1, K_h)$, may be chosen from among the approximate test functions $\tilde{v}$ and this leads to state: $\tilde{v} = \varphi_i$.

The approximate variational equation (1.110) is then written as:

$$\widetilde{(\mathbf{VP})} \left\{ \begin{array}{l} \text{Find } \tilde{u}_j, (j = 1, K_h) \text{ solution of:} \\ \displaystyle\sum_{j=1}^{K_h} \left(\int_\Omega \nabla \varphi_i \cdot \nabla \varphi_j \, d\Omega \right) \tilde{u}_j = \int_\Omega f \cdot \varphi_i \, d\Omega, \quad \forall i = 1, K_h . \end{array} \right. \tag{1.112}$$

Then the following is stated:

$$A_{ij} = \int_\Omega \nabla \varphi_i \cdot \nabla \varphi_j \, d\Omega \quad \text{and} \quad B_i = \int_\Omega f \cdot \varphi_i \, d\Omega . \tag{1.113}$$

Then, approximate variational problem $\widetilde{(\mathbf{VP})}$ is stated as:

$$\widetilde{(\mathbf{VP})} \left\{ \begin{array}{l} \text{Find } \tilde{u}_j, (j = 1, K_h), \text{ solution of:} \\ \displaystyle\sum_{j=1}^{K_h} A_{ij} \tilde{u}_j = B_i, \quad \forall i = 1, K_h . \end{array} \right. \tag{1.114}$$

This last form clearly shows the reduction that occurs as a result of the approximation process, by starting with variational problem $(\mathbf{VP})$, to give a problem having a finite dimension and whose resolution consists of a linear system of K_h equations with K_h unknowns.

The application of the finite elements method then necessitates the specification, in system (1.114), of the expression of basis functions φ_i, the operational method for the calculation of integrals intervening in the expression of the coefficients of matrix A_{ij} as well as that of the second member B_i and of course, the algorithm for the inversion of the linear system.

It is quite evident that, from a theoretical point of view it would be necessary to ensure that the matrix obtained is perfectly reversible before applying any inversion algorithm used on linear system (1.114) (see D.Euvrard, [5]).

Should the general problem of the approximation process that has been described at the beginning of this paragraph be reverted back to, it now appears that the successive chains of transformations that lead to the approximations that have been developed may be reassembled under the scheme below:

(Actual System) $\Rightarrow$ (Mathematical Model or Continuous Problem (**CP**)) ,

$\Rightarrow$ (Variational Formulation (**VP**)) ,

$\Rightarrow$ (Approximate Variational Problem $\widetilde{(\mathbf{VP})}$) ,

$\Rightarrow$ (Method for inversion of Linear System $\widetilde{\widetilde{(\mathbf{VP})}}$) .

The whole processing thus represented needs to really sensitise the numerical analyst engineer to show humbleness and care when publishing the final results.

It is true that theorems for estimating the error inherent to the method of finite elements do exist, but, as usual, this type of result cannot be global and concerns only part of the process demonstrated above.

In general, it relates to an evaluation of the error that appears following an approximation of variational problem (**VP**) by its approximate formulation $\widetilde{(\mathbf{VP})}$ (see for example, the Bramble-Hilbert lemma in D. Euvrard [5]).

Chapter 2
Some Fundamental Classes of Finite Elements

2.1 Variational Formulation and Approximations

Following the demonstration of the fundamental principles underlying the global methodology of the finite elements in Chapter 1, this chapter is dedicated to the approximation of variational formulations and to different choices generated by the finite elements method.

The whole process leads to the estimation of an approximate solution $\tilde{u}$ for a variational formulation (**VP**) as well as for a continuous problem (**CP**), both of which produce that form.

As seen previously (see 1.112), the Galerkin method is used to associate the variational formulation with the Laplace-Dirichlet problem and to give an approximate formulation ($\widetilde{\text{VP}}$) that is only a linear system needing to be reversed.

The resolution of this linear system offers approximation $\tilde{u}$ of the solution to variational problem (**VP**) and consequently an approximation of the solution to continuous problem (**CP**).

In fact, many mathematical models for engineering sciences lead to a formalism similar to the one demonstrated for the Laplace-Dirichlet problem.

A generic family of variational problems (**VP**) describing this formalism can be abstractly expressed in the form of:

$$(\textbf{VP}) \text{ Find } u \text{ belonging to } V \text{ solution of: } a(u,v) = L(v), \quad \forall v \in V, \qquad (2.1)$$

where:

- V is a vector space of functions,
- $a(.,.)$ is a bilinear form on $V \times V$,
- $L(.)$ is a linear form on V.

As shown in Chap. 1, additional investigations involving appropriate functional analysis techniques are essential to obtain a variational formulation (**VP**) having a unique solution equivalent to the solution of the continuous problem.

The approximation of the variational formulation (2.1) is essential in the wake of these conditions that "lock" the continuous and variational problems.

This state is closely related to the infinite dimension of functional spaces that emerge in most mathematical models applied to engineering sciences.

To this end, Galerkin suggests a method that consists in considering a subspace $\tilde{V}$, $(\tilde{V} \subset V)$, of finite dimension K_h that enables overcoming this incapacity of resolution of formulations having the same structure as formulation (2.1).

In this case, the abstract variational formulation (2.1) is transformed into the following approximation ($\widetilde{\text{VP}}$):

Find $\tilde{u}$ belonging to $\tilde{V}$ solution of:

$$a(\tilde{u}, \tilde{v}) = L(\tilde{v}), \quad \forall \tilde{v} \in \tilde{V}. \tag{2.2}$$

Since approximation space $\tilde{V}$ of finite dimension K_h has been introduced, it is now licit, and even natural, to consider a base of functions φ_i, $(i = 1, K_h)$, and to look for approximation $\tilde{u}$ that replaces solution u belonging to V in the form:

$$\tilde{u} = \sum_{j=1,K_h} \tilde{u}_j \varphi_j. \tag{2.3}$$

Note that decomposition (2.3) is fundamentally related to the shift of space V having an infinite dimension, to its internal approximation $\tilde{V}$ of finite dimension K_h.

Indeed, when considering a functional space V of infinite dimension, this form of decomposition (2.3) is no more relevant and the resolution of the variational problem (**VP**) stays whole except when considering particular vector spaces where any elements can be decomposed according to a basis composed of an infinite number of *denumerable* elements (as for separable Hilbert spaces).

In the approximate variational formulation ($\widetilde{\text{VP}}$) defined by (2.2), the specific choice of functions $\tilde{v}$ equal to basis functions φ_i, $(i = 1$ to $K_h)$, now allows rewriting formulation (2.2) as follows:

$$\left(\widetilde{\text{VP}}\right) \left[\begin{array}{l} \text{Find } \tilde{u} = {}^t[\tilde{u}_1, \ldots, \tilde{u}_{K_h}] \text{ belonging to } \tilde{V} \text{ solution of:} \\[2mm] a\left(\sum_{j=1,K_h} \tilde{u}_j \varphi_j, \varphi_i \right) = L(\varphi_i), \quad \forall i = 1 \text{ to } K_h. \end{array} \right. \tag{2.4}$$

The bilinear properties of form $a(.,.)$ and of the linear properties of form $L(.)$ are thus applied.

Therefore, the variational formulation $(\widetilde{\mathbf{VP}})$ is expressed in the form of:

$$\left(\widetilde{\mathbf{VP}}\right) \left[\begin{array}{l} \text{Find } \tilde{u} = {}^t[\tilde{u}_1, \dots, \tilde{u}_{K_h}] \text{ belonging to } \widetilde{V} \text{ solution of :} \\[2mm] \quad \sum_{j=1,K_h} a(\varphi_j, \varphi_i)\tilde{u}_j = L(\varphi_i), \quad \forall i = 1 \text{ to } K_h . \end{array} \right. \qquad (2.5)$$

The A_{ij} and b_i quantities are finally introduced and defined by:

$$A_{ij} = a(\varphi_j, \varphi_i), \quad b_i = L(\varphi_i) . \qquad (2.6)$$

From there, approximate variational formulation $(\widetilde{\mathbf{VP}})$ takes its following final form:

$$\left(\widetilde{\mathbf{VP}}\right) \left[\begin{array}{l} \text{Find } \tilde{u} = {}^t[\tilde{u}_1, \dots, \tilde{u}_{K_h}] \text{ belonging to } \widetilde{V} \text{ solution of :} \\[2mm] \quad \sum_{j=1,K_h} A_{ij}\tilde{u}_j = b_i, \quad \forall i = 1 \text{ to } K_h . \end{array} \right. \qquad (2.7)$$

As from then, it is noted that formulation (2.7) is none other than a linear system composed of matrix A of generic elements A_{ij} and of a second member b having component b_i.

Thus, it is established that any variational formulation (**VP**) expressed in form (2.1) and having form $a(.,.)$ and $L(.)$ are respectively bilinear and linear and can be solved by an approximation whose solution is equivalent to a linear system (2.7).

Of what nature is the data that can be defined within parameters and that is required to determine an effective solution to linear system (2.7) and consequently an approximation to variational formulation (2.1)?

Calculation of coefficients A_{ij} and that of second member b_i requires knowing basis functions $\varphi_i, (i = 1 \text{ to } K_h)$, of approximation space $\widetilde{V}$.

Of course, this knowledge closely depends on the definition of space $\widetilde{V}$ whose dimension K_h is finite.

This explains why one of the first ways to fix dimension K_h of space $\widetilde{V}$ consists in binding this dimension K_h to a finite number of values of functions $\tilde{v}$ belonging to $\widetilde{V}$ at privileged points or nodes $M_k, (k = 1 \text{ to } K)$, of integration domain Ω.

From then on, the introduction of an elementary geometry $G_m, (m = 1 \text{ to } M)$, is relevant since it generates a mesh of integration domain Ω, *de facto* providing the nodes of the geometrical discretisation (See Fig. 2.1).

These concepts gave rise to the Lagrange finite elements defined by triplet $(G, \Sigma, P(G))$ where:

- G defines the geometry of the elementary mesh (segment, triangle, square, polyhedron, etc.),

– $\Sigma = (M_1, \ldots, M_{K'})$, $(K' < K)$, denotes the group of nodes delimiting the elementary mesh G,

– $P(G)$ is the approximation space consisting of polynomials defined over G.

Finally, (G, Σ, P) triplet has to satisfy the unisolvence property defined as follows:

$$\forall (\xi_1, \ldots, \xi_{K'}) \in R^{K'}, \exists! p \in P(G)$$
$$\text{such that:} \quad p(M_k) = \xi_k, \quad \forall k = 1 \text{ to } K' \,. \tag{2.8}$$

In other words, there exists only one function p belonging to $P(G)$ going through the given K' values $(\xi_1, \ldots, \xi_{K'})$ at K' nodes delimiting elementary mesh G.

When the definition of functions belonging to $P(G)$ defined by a generating mesh G is known, the process of construction of approximation spaces $\tilde{V}$ within the framework of the Lagrange finite elements consists in stating:

$$\tilde{V} \equiv \left\{ \tilde{v} : \Omega \to \mathbf{R}, \tilde{v} \in C^0(\Omega), \tilde{v}|_G \in P(G) \right\}, \tag{2.9}$$

where the boundary conditions to which functions $\tilde{v}$ of $\tilde{V}$ may be subjected, depending on the problem considered, are disregarded.

Consequently, when disregarding boundary conditions that vary from one problem to another, the dimension of space $\tilde{V}$ defined by (2.9) is inferred from dimension K' of $P(G)$, from the number of mesh and from the number of nodes intervening in the geometrical discretisation of the integration domain Ω.

The generalisation of Lagrange finite elements is then done as follows:

Triplet $(G, \Sigma, P(G))$ defines a finite element with:

– An elementary mesh of geometrical discretisation G of $\mathbf{R}^n$, $(n = 1, 2 \text{ or } 3)$.

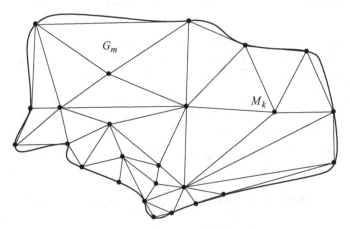

Fig. 2.1 Example of Mesh by Triangular Finite Elements

- A set Σ of degrees of freedom σ_k, $(k = 1$ to $K')$, consisting of linear forms on the space of defined functions on G.

- A vector space $P(G)$ of finite dimension equivalent to K'.

- *The unisolvance property*: For any K'-tuple of $\mathbf{R}^{K'}$ having real numbers, a unique element p exists and belongs to $P(G)$ such that:

$$\sigma_k(p) = \xi_k, \quad \forall k = 1 \text{ to } K' .$$

2.2 Convergence of the Finite Elements Method

As shown in paragraph 1.2.6 of Chap. 1, the significance of the different approximation levels resulting from the process of the modelling cascade of a numerical approximation, should encourage the numerical analyst to make a lucid and humble use of measuring tools which are necessary for the estimation of an approximation error in relation to the relevant numerical method.

In this case, the finite elements method offers a body of theoretical results which facilitates the estimation of the approximation error between solution u of variational problem (**VP**) and its approximation $\tilde{u}$ which is the solution to the approximate variational problem ($\widetilde{\textbf{VP}}$).

Considering the nature of the mathematical objects involved (functions u and $\tilde{u}$), a family of results are proposed in this paragraph to enable an estimation of the distance between solution u and its approximation $\tilde{u}$, ($\|u - \tilde{u}\|$), according to an *ad hoc* norm.

The family of variational problems (**VP**) having the abstract form below is used as an aid to the demonstration:

$$\text{Find } u \in V \text{ solution of: } a(u,v) = L(v), \quad \forall v \in V . \tag{2.10}$$

As in the previous paragraph, $\tilde{V}$ denotes the approximation space *internal* to V with finite dimension ($\tilde{V} \subset V$) having the generic element noted as $\tilde{v}$ and the approximate solution $\tilde{u}$ of solution u being a particular case approximation functions $\tilde{v}$ belonging to $\tilde{V}$.

In other words, approximate formulation ($\widetilde{\textbf{VP}}$) of variational problem (**VP**) is expressed as:

$$\text{Find } \tilde{u} \in \tilde{V} \text{ solution of: } a(\tilde{u},\tilde{v}) = L(\tilde{v}), \quad \forall \tilde{v} \in \tilde{V} . \tag{2.11}$$

Given the assumptions of the Lax-Milgram theorem 10 (See Chap. 1) associated with a Hilbertian norm noted as $\|.\|$, the following lemma is obtained:

▶ Lemma 4.

Variational Problem ($\widetilde{\textbf{VP}}$) (2.11) only admits one and only one solution $\tilde{u}$. Moreover, this solution satisfies the orthogonality relationship:

$$a(u - \tilde{u}, \tilde{v}) = 0, \quad \forall \tilde{v} \in \tilde{V} . \tag{2.12}$$

Proof: The existence and uniqueness of solution $\tilde{u}$ belonging to $\tilde{V}$ is immediate since the approximation is internal, $(\tilde{V} \subset V)$.

In fact, it is first noticed that the approximation space of finite dimension $\tilde{V}$ included in V is consequently a closed vector sub-space of V and thus presents a Hilbertian structure. The inclusion of $\tilde{V}$ in V allows the use of properties required for the application of a Lax-Milgram theorem to $\tilde{V}$.

As for the orthogonality relationship (2.12), the variational equation (**VP**) has to be expressed by substituting v by $\tilde{v}$:

$$a(u, \tilde{v}) = L(\tilde{v}), \quad \forall \tilde{v} \in \tilde{V} . \tag{2.13}$$

The difference between eqs. (2.13) and (2.11) immediately leads to orthogonality relationship (2.12).

The first result of the estimation of approximation error $\|u - \tilde{u}\|$ is known as the Céa's lemma.

▶ Lemma 5.

According to the hypothesis of Lax-Milgram theorem 10, if it is additionally assumed that approximation $\tilde{u}$ of exact solution u is internal, $(\tilde{V} \subset V)$, then the estimation of the error is obtained as:

$$\|u - \tilde{u}\| \leq C \inf_{\tilde{v} \in \tilde{V}} \|u - \tilde{v}\| . \tag{2.14}$$

Proof: The demonstration of Céa's lemma is based on the double control of quantity $a(u - \tilde{u}, u - \tilde{u})$ by using the property of V-ellipticity and the continuity of bilinear form $a(.,.)$.

For a start, according to orthogonality relationship (2.12), the following result is obtained by choosing $\tilde{v} = \tilde{u}$:

$$a(u - \tilde{u}, \tilde{u}) = 0 . \tag{2.15}$$

From then on, quantity $a(u - \tilde{u}, u - \tilde{u})$ may be expressed as:

$$\forall \tilde{v} \in \tilde{V} :$$
$$a(u - \tilde{u}, u - \tilde{u}) = a(u - \tilde{u}, u) - a(u - \tilde{u}, \tilde{u}) = a(u - \tilde{u}, u) \tag{2.16}$$
$$= a(u - \tilde{u}, u) - a(u - \tilde{u}, \tilde{v}) = a(u - \tilde{u}, u - \tilde{v}) . \tag{2.17}$$

Thus, by keeping the notations of the Lax-Milgram theorem 10, the following is obtained:

$$C_2 \|u - \tilde{u}\|^2 \leq a(u - \tilde{u}, u - \tilde{u}) = a(u - \tilde{u}, u - \tilde{v}) \leq C_1 \|u - \tilde{u}\| . \|u - \tilde{v}\| . \tag{2.18}$$

After simplification, the following is obtained:

$$\|u - \tilde{u}\| \leq \frac{C_1}{C_2} \|u - \tilde{v}\| , \quad \forall \tilde{v} \in \tilde{V} . \tag{2.19}$$

and expected constant C is none other than the ratio between C_1 and C_2.

The use of inequality control (2.19) is even more convincing considering that the upper bound of norm $\|u - \tilde{u}\|$ is minimised.

This explains how the result of Céa's lemma brings out the lower bound of quantities $\|u - \tilde{v}\|$ for any function $\tilde{v}$ belonging to $\tilde{V}$.

The next step consists in the characterisation of approximation space $\tilde{V}$ to determine the estimation of the error produced by Céa's lemma.

As mentioned in paragraph 2.1, Lagrange finite elements offer a simple solution for the systematic production of approximation space $\tilde{V}$ of finite dimension.

This process is based on the *unique* determination of an approximation function by considering its values taken at a finite number of points $M_k, (k = 1, K)$ situated on a given mesh of integration domain Ω.

From here, the reader needs to go back to the section detailing Lagrange finite elements (See paragraph 2.1) stating that the dimension of the approximation space $\tilde{V}$ corresponds to the number of nodes of the mesh of domain Ω if the boundary conditions that influence approximation functions $\tilde{v}$ are ignored.

This brings about the general introduction of the interpolation operator π_h defined as:

$$\pi_h : C^0(\bar{\Omega}) \rightarrow \tilde{V} \tag{2.20}$$
$$v \rightsquigarrow \pi_h v \equiv \sum_{k=1,K} v(M_k) \varphi_k ,$$

where φ_k denotes the basis function of approximation space $\tilde{V}$ characteristic of node M_k and satisfying the following property:

$$\varphi_k(M_l) = \delta_{kl}, (\delta_{kl} \text{ being the Krönecker symbol}) . \tag{2.21}$$

It is thus simple to verify that function $\pi_k v$, interpolated from v to K nodes M_k of the mesh of integration domain Ω is the unique function of $\tilde{V}$ proving:

$$\pi_h v(M_k) = v(M_k), \quad \forall k = 1, K. \tag{2.22}$$

It is thus licit to express the control inequality of Céa's lemma when specifically choosing $\tilde{v} = \pi_h u$:

$$\|u - \tilde{u}\| \leq C\|u - \tilde{v}\| = C\|u - \pi_h u\|. \tag{2.23}$$

Therefore, according to inequality control (2.23), the approximation error and the interpolation error are of the same order of magnitude.

This is why it suffices to estimate the interpolation error as a tool to measure the approximation error according to the nature and property of each Lagrange finite element.

The Bramble-Hilbert lemma is then introduced since it relies on these considerations to render the application of Céa's lemma fully operational.

In this present work, the demonstration is limited to the terms of the lemma having straight and unflattened finite elements and variational space V is considered to correspond to Sobolev space $H^1(\Omega)$.

Indeed, numerous problems arising from engineering sciences correspond to this functional framework (probably more regularly), knowing that, in any event, certain applications that may not fit in this framework would require mathematical techniques coming from a functional analysis that is way more than what this book can handle.

▶ **Lemma 6.**

Let h be the size of the elementary mesh of a Lagrange finite element. If approximation space $\tilde{V}$ *contains* the space of polynomials P_k having a degree less than or equal to k, in relation the pair of variables (x, y), then, for a finer discretisation and for any "sufficiently regular" solution u (at least in $H^1(\Omega)$) to variational problem (**VP**) of form (2.10), the following is obtained:

$$\|u - \pi_h u\|_{H^1(\Omega)} = O(h^k) \quad \text{and} \quad \|u - \tilde{u}\|_{H^1(\Omega)} = O(h^k). \tag{2.24}$$

Evidently, all the technicalities of the result of this lemma rest upon the estimation of the norm measuring the gap between solution u and its interpolation $\pi_h u$.

The preamble exposed using Céa's lemma was actually meant to underline the necessity of estimating this last norm in order to conclude on the approximation error of the finite elements method, at least in the context previously described.

2.3 Description of Ordinary Finite Elements

This section is dedicated to the introduction of the principal finite elements most commonly used in applications of engineering sciences.

In this chapter, each finite element is described in a systematic manner according to the following model:

1. The definition of elementary geometric mesh G,

2. The definition and dimension of approximation space $P(G)$

3. The definition of all linear forms σ_i on the space of functions defined on G.

4. The determination of the functions of the canonical basis of space $P(G)$, i.e. functions $(p_1, \ldots, p_{\dim P(G)})$ satisfying: $G_i(p_j) = \delta_{ij}$, where δ_{ij} denotes the Krönecker symbol.

▶ *Remark*

The existence of a collection of functions $(p_1, \ldots, p_{\dim P(G)})$ belonging to $P(G)$ satisfying canonical property $\sigma_i(p_j) = \delta_{ij}, \forall (i, j) \in \{1, \ldots, \dim P(G)\}$ implies that this system of function constitute a basis of $P(G)$.

In fact, it should be shown that system $(p_1, \ldots, p_{\dim P(G)})$ is an independent family in $P(G)$.

Let $(\alpha_1, \ldots, \alpha_{\dim P(G)}) \in R^{\dim P(G)}$ be, such that:

$$\sum_{i=1, \dim P(G)} \alpha_i p_i = 0. \tag{2.25}$$

Then show that linear combination (2.25) renders any coefficient α_i equal to zero.

To achieve this result, apply j^{th} linear form σ_j (for fixed j) to linear combination (2.25).

$$\sigma_j \left[\sum_{i=1, \dim P(G)} \alpha_i p_i \right] = \sigma_j(0). \tag{2.26}$$

The linear property of form σ_j along with the fact that $\sigma_j(0) = 0$ are then applied.

Equation (2.26) is then expressed as:

$$\sum_{i=1, \dim P(G)} \alpha_i \sigma_j(p_i) = \sum_{i=1, \dim P(G)} \alpha_i \delta_{ij} = \alpha_j = 0, \quad \forall j = 1, \ldots, \dim P(G). \tag{2.27}$$

The result shows that any coefficients α_j are all equal to zero and that $(p_1, \ldots, p_{\dim P(G)})$ is an independent family in a space of finite dimension $\dim P(G)$.

It is thus a generating family and consequently a basis to approximation space $P(G)$.

It will be noted that the "canonical" characterisation is relevant since each function p_i of this particular basis is characteristic of a privileged linear form σ_j considering that the other linear forms on canonical basis function p_i are equal to zero.

When particularly considering Lagrange finite elements, the linear forms show a number of particular values of the functions of $P(G)$ at certain points (or discretisation nodes) of the site of integration.

In this case, each function of the canonical basis corresponds to the unique function having value 1 at a given node of the discretisation and value 0 at the other nodes.

2.3.1 Finite Elements with a Space Variable

In this sub-paragraph, the finite elements presented are described by an elementary mesh consisting of the interval $G \equiv [0,1]$.

▶ Finite P_0 Element

1) Space $P(G) \equiv P_0$ is constituted by polynomials p being defined and constant on interval $[0,1]$.

The dimension of P_0 is obviously equal to 1.

2) Linear form σ is considered and is defined by:

$$\sigma : p \to \int_0^1 p(x)\,dx . \tag{2.28}$$

3) The unique function of the canonical basis for this element is the constant function equal to 1 on interval $[0,1]$.

To be sure, the definition of the function of the canonical basis is expressed according to the following agreed definition:

$$(\sigma(p) = 1) \Longleftrightarrow \left(\int_0^1 p(x)\,dx = 1 \right), \quad \text{where } p(x) = C^{te}, \quad \forall x \in [0,1] . \tag{2.29}$$

It is immediately deduced that

$$p(x) = 1, \quad \forall x \in [0,1] .$$

Functions $\tilde{v}$ belonging to $\tilde{V}$ are constant functions for each elementary mesh for this first finite element.

It would be noted that the constant on each mesh element corresponds to the mean value of function $\tilde{v}$ on the corresponding mesh.

▶ **Finite P_1 Element**

1) Approximation space $P(G) \equiv P_1$ consists of affine functions defined on elementary mesh $[0,1]$.

The dimension of space P_1 is equal to 2.

2) Both linear forms being considered are defined by:

$$\sigma_1 : p \to p(0), \quad \sigma_2 : p \to p(1). \tag{2.30}$$

3) In order to determine the functions of the canonical basis of space P_1 the property of both basis functions (p_1, p_2) are expressed as:

$$\left[\begin{array}{ll} \sigma_1(p_1) = 1 \Leftrightarrow p_1(0) = 1, & \sigma_1(p_2) = 0 \Leftrightarrow p_2(0) = 0, \\ \sigma_2(p_1) = 0 \Leftrightarrow p_1(1) = 0, & \sigma_2(p_2) = 1 \Leftrightarrow p_2(1) = 1. \end{array} \right. \tag{2.31}$$

It is then easily inferred that basis functions (p_1, p_2), solutions to (2.31) belonging to space P_1, consisting of defined affine functions on interval $[0,1]$, correspond to:

$$p_1(x) = 1 - x, \ p_2(x) = x. \tag{2.32}$$

▶ **Finite P_2 Elements**

1) Approximation space $P(G) \equiv P_2$ consists of polynomials having degrees less than or equal to two and defined on elementary mesh $[0,1]$.

The dimension of P_2 is equal to 3.

2) The three linear forms defined below are considered:

$$\sigma_1 : p \to p(0), \quad \sigma_2 : p \to p\left(\frac{1}{2}\right), \quad \sigma_3 : p \to p(1). \tag{2.33}$$

3) Now the properties of functions (p_1, p_2, p_3) of the canonical basis belonging to P_2 are expressed:

$$\left[\begin{array}{ll} \sigma_1(p_1) = 1 \Leftrightarrow p_1(0) = 1, & \sigma_1(p_2) = 0 \Leftrightarrow p_2(0) = 0, \\ \sigma_1(p_3) = 0 \Leftrightarrow p_3(0) = 0, & \sigma_2(p_1) = 0 \Leftrightarrow p_1(\frac{1}{2}) = 0, \\ \sigma_2(p_2) = 1 \Leftrightarrow p_2(\frac{1}{2}) = 1, & \sigma_2(p_3) = 0 \Leftrightarrow p_3(\frac{1}{2}) = 0, \\ \sigma_3(p_1) = 0 \Leftrightarrow p_1(1) = 0, & \sigma_3(p_2) = 0 \Leftrightarrow p_2(1) = 0, \\ \sigma_3(p_3) = 1 \Leftrightarrow p_3(1) = 1. \end{array} \right. \tag{2.34}$$

Then, use is made of the fact that each polynomial p_i of degree less than or equal to two is in the form of: $ax^2 + bx + c$.

The 9 coefficients of the 3 polynomials (p_1, p_2, p_3) are obtained from the 9 relationships (2.34).

The following is then obtained:

$$p_1(x) = (2x-1)(x-1), \quad p_2(x) = 4x(1-x), \quad p_3(x) = x(2x-1). \quad (2.35)$$

▶ Hermite's Finite Element

1) Approximation space $P(G) \equiv P_3$ consists of polynomials having degrees less than or equal to three and defined on elementary mesh $[0,1]$.

The dimension of P_3 is equal to 4.

2) Let the four linear forms be defined by:

$$\sigma_1 : p \to p(0), \quad \sigma_2 : p \to \frac{dp}{dx}(0), \quad \sigma_3 : p \to p(1), \quad \sigma_4 : p \to \frac{dp}{dx}(1). \quad (2.36)$$

3) The four functions (p_1, p_2, p_3, p_4) of canonical basis P_3 are determined. This result is achieved by expressing the sixteen relationships of the form $\sigma_i(p_j) = \delta_{ij}$:

$$\begin{bmatrix}
\sigma_1(p_1) = 1 \Leftrightarrow p_1(0) = 1, & \sigma_1(p_2) = 0 \Leftrightarrow p_2(0) = 0, \\
\sigma_1(p_3) = 0 \Leftrightarrow p_3(0) = 0, & \sigma_1(p_4) = 0 \Leftrightarrow p_4(0) = 0, \\
\sigma_2(p_1) = 0 \Leftrightarrow p_1'(0) = 0, & \sigma_2(p_2) = 1 \Leftrightarrow p_2'(0) = 1, \\
\sigma_2(p_3) = 0 \Leftrightarrow p_3'(0) = 0, & \sigma_2(p_4) = 0 \Leftrightarrow p_4'(0) = 0, \\
\sigma_3(p_1) = 0 \Leftrightarrow p_1(1) = 0, & \sigma_3(p_2) = 0 \Leftrightarrow p_2(1) = 0, \\
\sigma_3(p_3) = 1 \Leftrightarrow p_3(1) = 1, & \sigma_3(p_4) = 0 \Leftrightarrow p_4(1) = 0, \\
\sigma_4(p_1) = 0 \Leftrightarrow p_1'(1) = 0, & \sigma_4(p_2) = 0 \Leftrightarrow p_2'(1) = 0, \\
\sigma_4(p_3) = 0 \Leftrightarrow p_3'(1) = 0, & \sigma_4(p_4) = 1 \Leftrightarrow p_4'(1) = 1.
\end{bmatrix} \quad (2.37)$$

The sixteen relationships (2.37) again produce the sixteen coefficients of the four polynomials (p_1, p_2, p_3, p_4) of the canonical basis of P_3.

After a few calculations, the following is obtained:

$$\begin{aligned}
p_1(x) = (x-1)^2(2x+1), & \quad p_2(x) = x(x-1)^2, \\
p_3(x) = x^2(3-2x), & \quad p_4(x) = (x-1)x^2.
\end{aligned} \quad (2.38)$$

2.3.2 Finite Elements with Two Space Variables

2.3.2.1 Triangular Finite Elements

This section is dedicated to the introduction of finite elements whose elementary mesh G is a triangle with vertices M_1, M_2 and M_3 on plan $(O;x,y)$ (See Fig. 2.2).

▶ Finite P_0 Element

1) Approximation space $P(G) \equiv P_0$ consists of constant functions on triangle G.

The dimension of P_0 is equal to 1.

2) Linear form σ defined below is considered:

$$p \to \frac{1}{\text{Area}(G)} \iint_G p(x,y)\mathrm{d}x\mathrm{d}y. \tag{2.39}$$

3) Basis function p of P_0 satisfying property $\sigma(p) = 1$ is determined:

$$(\sigma(p) = 1) \Leftrightarrow \left(\frac{1}{\text{Area}(G)} \iint_G p(x,y)\mathrm{d}x\mathrm{d}y = 1 \right),$$

where:

$$p(x,y) = C^{te}, \quad \forall (x,y) \in G. \tag{2.40}$$

Then, the function of canonical basis p is the constant function equal to 1 on the whole of triangle G.

▶ Finite P_1 Element

1) Approximation space P_1 consists of polynomial functions having degrees less than or equal to one for the pair of variables (x,y).

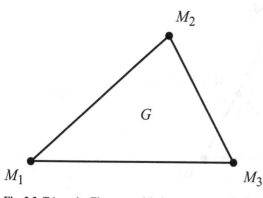

Fig. 2.2 Triangular Elementary Mesh

In other words, any function p of P_1 is expressed in the form:

$$p(x,y) = ax + by + c, \tag{2.41}$$

where (a,b,c) is a triplet of $\mathbf{R}^3$.

The previous definition leads to the conclusion that dimension P_1 is equal to 3.

2) The three linear forms defined below are considered:

$$\sigma_1 : p \rightarrow p(M_1), \quad \sigma_2 : p \rightarrow p(M_2), \quad \sigma_3 : p \rightarrow p(M_3). \tag{2.42}$$

3) The identification of the three functions of canonical basis (p_1, p_2, p_3) corresponds to the three barycentric functions $(\lambda_1, \lambda_2, \lambda_3)$ whose existence is established in the work of Daniel Euvrard [5].

Though, it is pointed out that, by definition, the polynomial functions of degree less than or equal to one for the pair (x, y) prove the canonical property:

$$\sigma_j(\lambda_i) \equiv \lambda_i(M_j) = \delta_{ij}. \tag{2.43}$$

▶ **Finite P_2 Element**

1) Approximation space $P(G) \equiv P_2$ consists of polynomial functions having degrees less or equal to 2 for the pair of variables (x, y).

In other words, any function p of P_2 is written in the form:

$$p(x,y) = ax^2 + by^2 + cxy + dx + ey + f, \tag{2.44}$$

where (a, b, c, d, e, f) is of any value and belongs to $\mathbf{R}^6$.

The previous definition (2.44) leads to the conclusion that the dimension of P_2 is equal to 6.

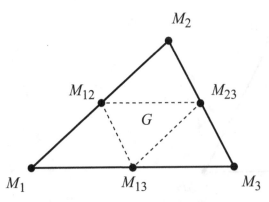

Fig. 2.3 Triangular Mesh for a Finite P_2 Element

2) To define the six linear forms $\sigma_i, (i = 1$ to $6)$, three new nodes (M_{12}, M_{13}, M_{23}) are introduced and placed in the middle of each side of triangle G (See Fig. 2.3).

It is then possible to introduce the six linear forms defined by:

$$\sigma_1: p \to p(M_1), \qquad \sigma_2: p \to p(M_2), \tag{2.45}$$

$$\sigma_3: p \to p(M_3), \qquad \sigma_4: p \to p(M_{12}), \tag{2.46}$$

$$\sigma_5: p \to p(M_{13}), \qquad \sigma_6: p \to p(M_{23}). \tag{2.47}$$

3) The construction of the functions of canonical basis $(p_1, p_2, p_3, p_4, p_5, p_6)$ is worked out in the following way:

Function p_1 may be taken as an example. This second degree polynomial in relation to the pair (x, y) must be zero at the points below: M_2, M_3, M_{12}, M_{13} and M_{23}.

Therefore, polynomial p_1, whose trace is a second degree trinomial of the oblique variable defining the parameter of segment M_2M_3, is identically zero over segment M_2M_3, being zero at the three points M_2, M_3 and M_{23}.

Moreover, as segment M_2M_3 is characterised by equation $\lambda_1 = 0$, it means that λ_1 can be factorised in the polynomial expression p_1.

In the same way, polynomial p_1 is zero at nodes M_{13} and M_{12}. Since the barycentric functions λ_i are affine in x and in y, in these two nodes, λ_1 is exactly equal to $1/2$ on segment $M_{13}M_{12}$.

In other words, by factorising p_1 by the quantity $\lambda_1 - \frac{1}{2}$, it is ensured that p_1 is really zero at nodes M_{13} and M_{12}.

Therefore, the polynomial structure of function p_1 is written as:

$$p_1(M) = \alpha \lambda_1(M) \left(\lambda_1(M) - \frac{1}{2} \right), \tag{2.48}$$

where α is a constant which must be determined so that polynomial p_1 may equal 1 at its characteristic node, namely at node M_1.

Moreover, it can be noted that expression (2.48) indeed confers a second degree polynomial structure in relation to the pair of variables (x, y) to function p_1 because polynomial λ_1 is of the first degree in relation to the pair (x, y).

The following is then written as:

$$p_1(M_1) \equiv \alpha \lambda_1(M_1) \left(\lambda_1(M_1) - \frac{1}{2} \right) = \alpha \times \frac{1}{2}. \tag{2.49}$$

To ensure the property $p_1(M_1) = 1$, the value of coefficient α can then be inferred therefrom: $\alpha = 2$.

Polynomial p_1 is finally written as:

$$p_1(M) = \lambda_1(M)(2\lambda_1(M) - 1). \tag{2.50}$$

The other polynomials of the canonical basis may be inferred by the same method and the following is obtained:

$$p_1(M) = \lambda_1(M)(2\lambda_1(M) - 1), \quad p_2(M) = \lambda_2(M)(2\lambda_2(M) - 1), \quad (2.51)$$
$$p_3(M) = \lambda_3(M)(2\lambda_3(M) - 1), \quad p_{12}(M) = 4\lambda_1\lambda_2(M), \quad (2.52)$$
$$p_{13}(M) = 4\lambda_1\lambda_3(M), \quad p_{23}(M) = 4\lambda_2\lambda_3(M). \quad (2.53)$$

▶ Finite P_3 Elements

1) The approximation space $P(G) \equiv P_3$ consists of polynomial functions of degree less or equal to three for the pair of variables (x, y).

In other words, any function p of P_3 is written in the form:

$$p(x, y) = ax^3 + by^3 + cx^2y + dxy^2 + ex^2 + fy^2 + gxy + hx + iy + j, \quad (2.54)$$

where $(a, b, c, d, e, f, g, h, i, j)$ is of any value and belongs to $\mathbf{R}^{10}$.

The previous definition (2.54) leads to the conclusion that the dimension of P_3 is equal to 10.

2) To define the ten linear forms $\sigma_i, (i = 1 \text{ to } 10)$, seven new nodes $(M_{112}, M_{122}, M_{113}, M_{133}, M_{223}, M_{233}, M_{123})$ are introduced and placed in the third part of each side of triangle G, (See Fig. 2.4).

It is then possible to introduce the ten linear forms by using:

$$\sigma_1: p \to p(M_1), \quad \sigma_2: p \to p(M_2), \quad (2.55)$$
$$\sigma_3: p \to p(M_3), \quad \sigma_4: p \to p(M_{112}), \quad (2.56)$$
$$\sigma_5: p \to p(M_{122}), \quad \sigma_6: p \to p(M_{223}), \quad (2.57)$$
$$\sigma_7: p \to p(M_{233}), \quad \sigma_8: p \to p(M_{113}), \quad (2.58)$$
$$\sigma_9: p \to p(M_{133}), \quad \sigma_{10}: p \to p(M_{123}). \quad (2.59)$$

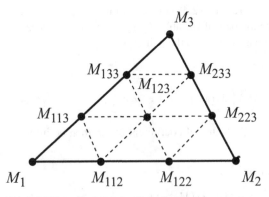

Fig. 2.4 Triangular Mesh for Finite P_3 Element

3) The determination of the ten functions of canonical basis $(p_1, (i = 1 \text{ to } 10))$ is worked out by using the same principles presented for the triangular finite $\mathbf{P_2}$ elements.

The case of polynomial p_1, which is characteristic of the node, M_1 can be studied again, since it satisfies the property: $p_1(M_1) = 1$.

Polynomial p_1 being zero at the other nine nodes, the following factorisations may be inferred:

- λ_1 is factorised in the expression of p_1, since this polynomial must be zero at nodes $(M_2, M_3, M_{223}, M_{233})$.

- $(\lambda_1 - 2/3)$ is factorised in the expression of p_1, since this polynomial must be zero at nodes (M_{112}, M_{113}).

- $(\lambda_1 - 1/3)$ is factorised in the expression of p_1, since this polynomial must be zero at nodes $(M_{122}, M_{133}, M_{123})$.

Therefore, polynomial p_1 takes the following form:

$$p_1(M) = \alpha \lambda_1(M) \left(\lambda_1(M) - \frac{1}{3} \right) \left(\lambda_1(M) - \frac{2}{3} \right), \tag{2.60}$$

where, once again, constant α must be adjusted so that polynomial p_1 is equal to 1 at node M_1.

Moreover, it would be noted that the shape of polynomial p_1 (2.60) is coherent with that of definition (2.54) of the functions belonging to P_3 in accordance with the fact that the barycentric function λ_1 is a first degree polynomial in relation to the pair of variables (x, y).

The following is then easily obtained: $\alpha = \frac{9}{2}$ and the final shape of polynomial p_1 is thus:

$$p_1(M) = \frac{9}{2}\lambda_1(M) \left(\lambda_1(M) - \frac{1}{3} \right) \left(\lambda_1(M) - \frac{2}{3} \right). \tag{2.61}$$

Polynomials p_2 and p_3 are immediately inferred from the expression of polynomial p_1, for reasons of obvious symmetry:

$$p_2(M) = \frac{9}{2}\lambda_2(M) \left(\lambda_2(M) - \frac{1}{3} \right) \left(\lambda_2(M) - \frac{2}{3} \right), \tag{2.62}$$

$$p_3(M) = \frac{9}{2}\lambda_3(M) \left(\lambda_3(M) - \frac{1}{3} \right) \left(\lambda_3(M) - \frac{2}{3} \right). \tag{2.63}$$

Polynomial p_{112} is now studied. This polynomial presents the following factorisations:

- λ_1 is factorised in the expression of p_{112}, since this polynomial must be zero at nodes $(M_2, M_3, M_{223}, M_{233})$.

- λ_2 is factorised in the expression of p_{112}, since this polynomial must be zero at nodes $(M_1, M_3, M_{113}, M_{133})$.

- $(\lambda_1 - 1/3)$ is factorised in the expression of p_1, since this polynomial must be zero at nodes $(M_{122}, M_{133}, M_{123})$.

Hence, the structure of p_{112} is given by:

$$p_{112}(M) = \beta \lambda_1(M) \lambda_2(M) \left(\lambda_1(M) - \frac{1}{3} \right) , \qquad (2.64)$$

where the constant β must be adjusted so that the polynomial p_{112} is equal to one at node M_{112}.

By therefore writing that $\lambda_1 = 2/3$ and $\lambda_2 = 1/3$ at node M_{112}, the following is hence obtained:

$$\beta = \frac{27}{2} . \qquad (2.65)$$

The basis function p_{112} is finally written as:

$$p_{112}(M) = \frac{27}{2} \lambda_1(M) \lambda_2(M) \left(\lambda_1(M) - \frac{1}{3} \right) . \qquad (2.66)$$

Once again, for reasons of symmetry, the other basis functions p_{ijk}, where (i, j, k) differs from triplet $(1, 2, 3)$, are written as:

$$p_{122}(M) = \frac{27}{2} \lambda_1(M) \lambda_2(M) \left(\lambda_2(M) - \frac{1}{3} \right) , \qquad (2.67)$$

$$p_{113}(M) = \frac{27}{2} \lambda_1(M) \lambda_3(M) \left(\lambda_1(M) - \frac{1}{3} \right) , \qquad (2.68)$$

$$p_{133}(M) = \frac{27}{2} \lambda_1(M) \lambda_3(M) \left(\lambda_3(M) - \frac{1}{3} \right) , \qquad (2.69)$$

$$p_{223}(M) = \frac{27}{2} \lambda_2(M) \lambda_3(M) \left(\lambda_2(M) - \frac{1}{3} \right) , \qquad (2.70)$$

$$p_{233}(M) = \frac{27}{2} \lambda_1(M) \lambda_3(M) \left(\lambda_3(M) - \frac{1}{3} \right) . \qquad (2.71)$$

This study can be concluded by the analysis of the last polynomial function p_{123} of the canonical basis of P_3.

This polynomial presents the following factorisations:

- λ_1 is factorised in the expression of p_{123}, since this polynomial must be zero at nodes $(M_2, M_3, M_{223}, M_{233})$.

- λ_2 is factorised in the expression of p_{123}, since this polynomial must be zero at nodes $(M_1, M_3, M_{113}, M_{133})$.

- λ_3 is factorised in the expression of p_{123}, since this polynomial must be zero at nodes $(M_1, M_2, M_{112}, M_{122})$.

Hence, function p_{123} possesses the following polynomial structure:

$$p_{123}(M) = \gamma \lambda_1(M) \lambda_2(M) \lambda_3(M) , \tag{2.72}$$

where constant γ is adjusted so that the polynomial p_{123} may satisfy its characteristic property at node M_{123}, namely: $p_{123}(M_{123}) = 1$.

Considering that the barycentric functions λ_1, λ_2 and λ_3 have, all three, the same value of $1/3$ at node M_{123}, the constant γ is then equal to:

$$\gamma = 27 . \tag{2.73}$$

The polynomial p_{123} is finally written as:

$$p_{123}(M) = 27 \lambda_1(M) \lambda_2(M) \lambda_3(M) . \tag{2.74}$$

2.3.2.2 Quadrangular Finite Elements

This section is dedicated to the presentation of finite elements, whose elementary mesh G is the square $[0,1] \times [0,1]$ of vertices M_1, M_2, M_3 and M_4 in the plan $(O; x, y)$, (See Fig. 2.5).

▶ Finite Q_1 Element

1) Space $P(G) \equiv Q_1$ is defined as comprising the set of the polynomials of degree less or equal to 1 for each of the variables x and y.

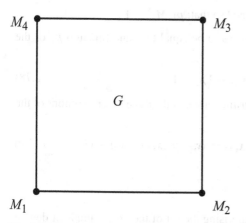

Fig. 2.5 Elementary Square Mesh for Plane Finite Element

Hence, any function p belonging to Q_1 is written as:

$$p(x,y) = axy + bx + cy + d,\qquad(2.75)$$

where (a, b, c, d) represents any value and describes $\mathbf{R}^4$.

A simple examination of definition (2.75) confirms that the dimension of Q_1 is equal to 4.

2) The four linear forms which are then introduced are defined by:

$$\sigma_i : p \rightarrow p(M_i),\quad \forall i = 1 \text{ to } 4.\qquad(2.76)$$

3) To determine the four functions $p_i, (i = 1 \text{ to } 4)$ of the canonical basis of space Q_1, it is noted that these functions must satisfy the definition:

$$\sigma_j(p_i) = p_i(M_j) = \delta_{ij}.$$

Therefore, each function of the canonical basis is characteristic of a unique vertice which defines the square G, that is, by taking the value of 1 at this characteristic vertice and 0 at the three other vertices.

This is why the following factorisation properties are identified.

The properties of polynomial p_1 are described below:

- p_1 having a zero value on segment M_2M_3 defined by $x = 1$, the monomial $(x-1)$ must be factorised in the expression of p_1.

- p_1 having a zero value on segment M_3M_4 parameterised by $y = 1$, the monomial $(y-1)$ must be factorised in the expression of p_1.

Therefore, the structure of the function of the canonical basis p_1 is:

$$p_1(x,y) = \alpha(x-1)(y-1),\qquad(2.77)$$

where, as usual, constant α is determined so that: $p_1(M_1) = 1$.

Coefficient α is then worked out so as to be equal to 1 and function p_1 of the canonical basis is written as:

$$p_1(x,y) = (x-1)(y-1).\qquad(2.78)$$

An analogous reasoning enables the writing down of the three other functions of the canonical basis of Q_1 :

$$p_2(x,y) = x(1-y),\quad p_3(x,y) = xy,\quad p_4(x,y) = y(1-x).\qquad(2.79)$$

▶ **Finite Q_2 Element**

1) Space $P(G) \equiv Q_2$ is defined as comprising the set of the polynomials of degree less or equal to 2 for each of the variables x and y.

Hence, any function p belonging to Q_2 is written as:

$$p(x,y) = ax^2y^2 + bx^2y + cxy^2 + dx^2 + ey^2 + fxy + gx + hy + i,\qquad(2.80)$$

where $(a, b, c, d, e, f, g, h, i)$ represents any value and describes $\mathbf{R}^9$.

Definition (2.80) therefore implies that the dimension of Q_2 is equal to 9.

2) To define the nine linear forms of σ_i, five additional discretisation nodes M_5, M_6, M_7, M_8 and M_9 are introduced, so that the first four nodes correspond to the middle of each side of square G, while M_9 indicates the centre of the square (See Fig. 2.6).

The nine linear forms $\sigma_i, (i = 1$ to $9)$ are then defined by:

$$\sigma_i : p \rightarrow p(M_i), \quad \forall i = 1 \text{ to } 9.\qquad(2.81)$$

3) The same construction process, presented in the case of the quadrangular finite $\mathbf{Q_1}$ element, is strictly applied and the nine functions of the canonical basis $p_i, (i = 1$ to $9)$ is easily obtained:

$$
\begin{aligned}
p_1(x,y) &= (1-x)(1-2x)(1-y)(1-2y),\\
p_2(x,y) &= x(2x-1)(1-y)(1-2y),\\
p_3(x,y) &= xy(2x-1)(2y-1),\\
p_4(x,y) &= y(1-x)(1-2x)(2y-1),\\
p_5(x,y) &= 4x(1-x))(1-y)(1-2y),\\
p_6(x,y) &= 4xy(2x-1)(1-y),\\
p_7(x,y) &= 4xy(1-x)(2y-1),\\
p_8(x,y) &= 4y(1-x)(1-2x)(1-y),\\
p_9(x,y) &= 16xy(1-x)(1-y).
\end{aligned}
\qquad(2.82)
$$

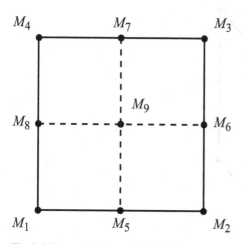

Fig. 2.6 Elementary Square Mesh for Finite $\mathbf{Q_2}$ Element

2.3.3 *Finite Elements with Three Space Variables*

▶ Cubic Finite Element

This last section is devoted to the presentation of a finite element, whose elementary mesh G is the cube $[0,1] \times [0,1] \times [0,1]$ of vertices $M_i, (i = 1$ to $8)$, in the space $(O;x,y,z)$ (See Fig. 2.7).

1) Space Q_1 is defined as comprising the set of the polynomials of degree less or equal to one for each of variables x, y and z.

Therefore any function p belonging to Q_1 is written as:

$$p(x,y,z) = axyz + bxy + cxz + dyz + ex + fy + gz + h, \qquad (2.83)$$

where (a, b, c, d, e, f, g, h) represents any value and describes $\mathbf{R}^8$.

The examination of definition (2.83) clearly shows that the dimension of Q_1 is equal to 8.

2) The eight linear forms which are then introduced are defined by:

$$\sigma_i : p \rightarrow p(M_i), \quad \forall i = 1 \text{ to } 8 . \qquad (2.84)$$

3) The eight functions of the canonical basis $p_i, (i = 1$ to $8)$ can constructed at present, that is, by satisfying: $\sigma_j(p_i) \equiv p_i(M_j) = \delta_{ij}$.

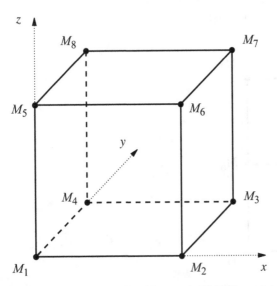

Fig. 2.7 Cubic Mesh for Finite Element in Three Dimensional Space

To proceed, it is sufficient to construct each of these eight polynomial functions p_i, by identifying the monomials that need to be factorised in the expression of each of the polynomial functions.

The polynomial p_1 which is characteristic of node M_1 and which is zero at the other seven nodes M_i is considered.

a) The monomial $(1-x)$ is factorised in the expression of p_1 so that it takes a zero value at nodes M_2, M_3, M_6 and M_7.

b) The monomial $(1-y)$ is factorised in the expression of p_1 so that it takes a zero value at nodes M_3, M_4, M_7 and M_8.

c) The monomial $(1-z)$ is factorised in the expression of p_1 so that it takes a zero value at nodes M_5, M_6, M_7 and M_8 .

Therefore, the basis function p_1 presents the following structure:

$$p(x,y,z) = \alpha(1-x)(1-y)(1-z), \qquad (2.85)$$

where constant α is determined so as to adjust the value of polynomial p_1 to 1 within its characteristic node M_1.

However, at node M_1, $x = y = z = 0$ implies that constant α equals 1 and function p_1 is finally written as:

$$p(x,y,z) = (1-x)(1-y)(1-z). \qquad (2.86)$$

An analogous reasoning enables the hop by hop formation of each polynomial of the canonical basis of Q_1.

The basis, thus constructed, is written as:

$$
\begin{aligned}
&p_1(x,y,z) = (1-x)(1-y)(1-z), \quad &&p_2(x,y,z) = x(1-y)(1-z), \\
&p_3(x,y,z) = xy(1-z), \quad &&p_4(x,y,z) = (1-x)y(1-z), \\
&p_5(x,y,z) = (1-x)(1-y)z, \quad &&p_6(x,y,z) = x(1-y)z, \\
&p_7(x,y,z) = xyz, \quad &&p_8(x,y,z) = (1-x)yz.
\end{aligned}
\qquad (2.87)
$$

Chapter 3
Variational Formulations

The purpose of this chapter is to develop and analyse the variational formulations for a one-dimensional case, on the one hand and to apply the $\mathbf{P_1}$ elements of Lagrange, on the other hand.

In fact, in accordance with the degree of complexity of the problems tackled, it will be suggested to apply the following program of analysis, which is common to all partial differential equations, whose approximation is sought by the finite elements method:

- **Step A**: This first step aims at obtaining a variational formulation (**VP**) (or weak formulation) for a given continuous problem (**CP**) (or strong formulation).

- **Step B**: The existence and uniqueness of weak solutions of the problem (**VP**) are studied, (essentially the application of the Lax-Milgram theorem).

- **Step C**: This step is devoted to the analysis of the regularity of the weak solutions of the variational problem (**VP**).

- **Step D**: This part deals with the equivalence of strong and weak formulations. It is particularly shown that a weak solution of the variational problem (**VP**), which moreover shows the regularity property as obtained in Step **C**, is a strong solution of the continuous problem (**CP**).

- **Step E**: This step is devoted to the writing of the nodal equation system that provides an approximation of the weak solution of the variational problem (**VP**).

Furthermore, nodal equations obtained will be compared to those obtained with the finite differences scheme that are associated with the continuous problem (**CP**).

Hence, two distinct parts may be studied separately. A first theoretical part will be devoted to the analysis of the existence, uniqueness and regularity of the solutions to variational formulations as well as to the notion of equivalence between continuous and variational problems (Steps **A**, **B**, **C** and **D**).

The second part, which is totally different, will be devoted to obtaining an approximation of the different nodal equations, mainly using the Lagrange finite element P_1 and to the analysis of some schemes having finite differences.

Therefore, the reader may, at leisure, deal with the whole aspect of the problem (both theoretical and numerical) or study either of these parts.

However, in the case where only the numerical part needs to be studied, it would be suitable to refer to the theoretical part, or at least the first question, so as to elaborate and determine the proper variational formulation for the numerical application of finite elements.

3.1 Dirichlet's Problem

3.1.1 Statement

The aim of this problem is to propose a mathematical and numerical study of the solution to a linear differential equation subjected to Dirichlet boundary conditions.

Find $u \in H^2(0,1)$ being the solution to:

$$(\mathbf{CP}) \begin{cases} -u''(x) + u(x) = f(x), \ 0 \le x \le 1, \\ u(0) = u(1) = 0, \end{cases} \qquad (3.1)$$

in which f is a given function belonging to $L^2(0,1)$.

Besides, it is pointed out that Sobolev's space $H^2(0,1)$ is defined as:

$$H^2(0,1) = \left\{ v :]0,1[\to \mathbf{R}, \frac{\mathrm{d}^k v}{\mathrm{d}x^k} \in L^2(0,1), \ \forall k = 0,1,2 \right\}. \qquad (3.2)$$

▶ **Variational Formulation – Theoretical Part**

1) Let v be a test function, defined from $[0,1]$ to $\mathbf{R}$, belonging to a functional space V whose characteristics will be determined *a posteriori*.

Show that the continuous problem (**CP**) may be expressed in a variational formulation (**VP**) in the form:

$$a(u,v) = L(v), \quad \forall v \in V.$$

The bilinear form $a(.,.)$, the linear form $L(.)$ as well as the functional space V need to be specified.

2) Establish the existence and uniqueness of the weak solution of the variational problem (VP) in $H_0^1(0,1)$, in which $H_0^1(0,1)$ is defined as:

$$H_0^1(0,1) = \left\{ v :]0,1[\to \mathbf{R}, \ v \text{ and } v' \in L^2(0,1), \ v(0) = v(1) = 0 \right\} . \quad (3.3)$$

3) Show that any weak solution of the variational problem (VP) also belongs to $H^2(0,1)$.

4) To infer the equivalence between the strong formulation (CP) set in $H^2(0,1)$ and the weak formulation (VP) considered in:

$$H_0^1(0,1) \cap H^2(0,1) .$$

▶ **Numerical Part – Lagrange Finite P_1 Elements**

5) The approximation of the variational problem (VP) is worked out using the Lagrange finite elements P_1.

This is performed by introducing a regular mesh of $[0,1]$ interval, of constant step h, such as:

$$\begin{cases} x_0 = 0, \quad x_{N+1} = 1 , \\ x_{i+1} = x_i + h, \ i = 0 \text{ to } N . \end{cases} \quad (3.4)$$

The approximation space $\tilde{V}$ can now be defined as:

$$\tilde{V} = \left\{ \tilde{v} / \tilde{v} \in C^0([0,1]), \ \tilde{v}|_{[x_i,x_{i+1}]} \in P_1, \ \tilde{v}(0) = \tilde{v}(1) = 0 \right\} , \quad (3.5)$$

in which $P_1 \equiv P_1([x_i,x_{i+1}])$ refers to the polynomial space which is defined over $[x_i,x_{i+1}]$, having a degree less or equal to one.

– What is the dimension of $\tilde{V}$?

6) Let $\varphi_i, (i = 1 \text{ to } \dim\tilde{V})$, be the canonical basis of $\tilde{V}$ establishing $\varphi_i(x_j) = \delta_{ij}$, in which δ_{ij} refers to the Krönecker symbol.

After having written the approximate variational formulation $(\widetilde{VP})$, of solution $\tilde{u}$, which is associated to the variational problem (VP), show that by choosing:

$$\tilde{v}(x) = \varphi_i(x), \ (i = 1 \text{ to } \dim\tilde{V}) \quad \text{and} \quad \tilde{u}(x) = \sum_{j=1,\dim\tilde{V}} \tilde{u}_j \varphi_j , \quad (3.6)$$

the following system $(\widetilde{VP})$ is obtained:

$$(\widetilde{PV}) \sum_{j=1,\dim\tilde{V}} A_{ij}\tilde{u}_j = b_i, \quad \forall i \in \{1,\ldots,\dim\tilde{V}\} , \quad (3.7)$$

where it has been observed that:

$$A_{ij} = \int_0^1 (\varphi_i' \varphi_j' + \varphi_i \varphi_j) dx, \quad b_i = \int_0^1 f \varphi_i dx. \qquad (3.8)$$

▶ **Function φ_i Characteristic of a Node Strictly Interior at [0,1]**

7) Given the regularity of the mesh, the generic nodal equation of the $(\widetilde{\mathbf{VP}})$ system associated to any basis function φ_i, which is characteristic of a node strictly interior at $[0,1]$, is expressed as:

$$(\widetilde{\mathbf{VP}}_{\text{Int}}) A_{i,i-1} \tilde{u}_{i-1} + A_{i,i} \tilde{u}_i + A_{i,i+1} \tilde{u}_{i+1} = b_i, \quad (\forall i = 1 \text{ to } K_h), \qquad (3.9)$$

where it has been assumed that:

$$K_h = \dim \tilde{V}.$$

– Using the trapezium formula, calculate the 4 coefficients of (A_{ij}, b_i).

8) Group the results together by writing down the corresponding nodal equation.

9) Show that the centrered finite differences scheme associated with the differential equation of the continuous problem **(CP)** is obtained again. What is its degree of precision?

It is pointed out that the trapezium quadrature formula is written as:

$$\int_a^b \xi(s) ds \simeq \frac{(b-a)}{2} \{\xi(a) + \xi(b)\}.$$

3.1.2 Solution

▶ **Variational Formulation – Theoretical Part**

A.1) Let v be a test function, defined on $[0,1]$ having real values and "sufficiently regular". Each time a variational formulation is needed, the regularity of the functions v will be specified *a posteriori*, so that the formulation is significant enough to be understood.

The differential equation of the continuous problem (**CP**) is then multiplied by v and is integrated along the interval $[0,1]$.

$$-\int_0^1 u''v\,dx + \int_0^1 uv\,dx = \int_0^1 fv\,dx, \quad \forall v \in V. \tag{3.10}$$

An integration by parts moreover leads to:

$$\int_0^1 u'v'\,dx + u'(0)v(0) - u'(1)v(1) + \int_0^1 uv\,dx = \int_0^1 fv\,dx, \quad \forall v \in V. \tag{3.11}$$

It is now observed that the homogeneous boundary conditions for u, $(u(0) = u(1) = 0)$ do not appear in the integral formulation (3.11).

In order to retain the whole information of the continuous problem (**CP**) in the future variational formulation (**VP**), it would therefore be suitable to impose that test functions v fulfil the boundary conditions:

$$v(0) = v(1) = 0. \tag{3.12}$$

Such a method indeed ensures that the solution u, as one of the functions v of the searched variational space V, will have all the properties required at the boundary conditions on $[0,1]$.

The following formal variational formulation is thus obtained:

Find u belonging to V being the solution of:

$$\int_0^1 (u'v' + uv)\,dx = \int_0^1 fv\,dx, \quad \forall v \text{ such that: } v(0) = v(1) = 0. \tag{3.13}$$

In fact, this variational formulation is indeed formal, since it is necessary to specify the regularity of the test functions, which enables the equ. (3.13) to acquire significance, especially the convergence of the integrals of the equation.

This consequently leads to the specification of the functional space V within which the solution u of the integral formulation (3.13) would be found out.

This is performed by making use of the Cauchy-Schwartz Inequality that produces the following inequality of control:

$$\left| \int_0^1 u'v'\,dx \right| \leq \int_0^1 |u'v'|\,dx \leq \left[\int_0^1 |u'|^2 dx \right]^{1/2} \cdot \left[\int_0^1 |v'|^2 dx \right]^{1/2}, \tag{3.14}$$

$$\left|\int_0^1 fv\mathrm{d}x\right| \le \int_0^1 |fv|\mathrm{d}x \le \left[\int_0^1 |f|^2\mathrm{d}x\right]^{1/2} \cdot \left[\int_0^1 |v|^2\mathrm{d}x\right]^{1/2}. \qquad (3.15)$$

Given that the inequality (3.14) can be rewritten by substituting u' by u and v' by v, so as to process the integral bearing on the product uv by the same method.

Therefore, if the variational space V is determined as being the set of the functions v belonging to $L^2(0,1)$ and whose first derivative also belongs to $L^2(0,1)$, the variational equation (3.13) is correctly defined.

To conclude, and by adding the homogenous Dirichlet boundary conditions (3.12), the variational space V, in which the solution u of the variational formulation (**VP**) will be sought is nothing else but the Sobolev space $H_0^1(0,1)$, which is defined by (3.3). Finally, the variational formulation (**VP**) is written as:

$$(\textbf{VP}) \begin{cases} \text{Find } u \text{ belonging to } V \text{ being the solution of: } a(u,v) = L(v), \ \forall v \in V, \text{ where:} \\[2mm] a(u,v) \equiv \int_0^1 \left[u'(x)v'(x) + u(x)v(x)\right]\mathrm{d}x\,, \\[2mm] L(v) \equiv \int_0^1 f(x)v(x)\mathrm{d}x\,, \\[2mm] V \equiv H_0^1(0,1)\,. \end{cases} \qquad (3.16)$$

A.2) In order to prove the existence and uniqueness of the solution pertaining to the variational problem (**VP**) (3.16), the application of Lax-Milgram theorem 4 requires the choice of a norm to be defined in the functional space $H_0^1(0,1)$.

Yet, as $H_0^1(0,1) \subset H^1(0,1)$, it is natural to evaluate the size of the functions of $H_0^1(0,1)$ using the natural norm of $H^1(0,1)$.

In other words, the following formula is proposed:

$$\forall v \in H_0^1(0,1): \|v\|_{H^1}^2 \equiv \int_0^1 v^2\mathrm{d}x + \int_0^1 v'^2\mathrm{d}x \equiv \|v\|_{L^2}^2 + \|v'\|_{L^2}^2. \qquad (3.17)$$

It is then seen that the bilinear form $a(.,.)$ is none other than the inner product from which the H^1 norm (3.17) is obtained:

$$\forall v \in H^1(0,1): a(v,v) \equiv (v,v)_{H^1} = \|v\|_{H^1}^2\,,$$

in which $(.,.)_{H^1}$ has been written as the inner product in H^1.

Under these conditions, the Sobolev spaces $H^1(0,1)$ and $H_0^1(0,1)$, (as a subspace of $H^1(0,1)$, also is a Hilbert space for the norm (3.17)), (the works of P. A. Raviart [7] or H. Brézis [1] may be referred to for clarifications).

Moreover, the continuity constant for the bilinear form $a(.,.)$ can be then easily be obtained since it only necessary to use the Cauchy-Schwartz Inequality to find the inner product $(.,.)_{H^1}$:

$$|a(u,v)| \equiv |(u,v)_{H^1}| \le (u,u)_{H^1}^{1/2} \cdot (v,v)_{H^1}^{1/2} = \|u\|_{H^1} \cdot \|v\|_{H^1}. \qquad (3.18)$$

In other words, the continuity constant for the form $a(.,.)$ is equal to one.

The continuity for the linear form L is furthermore obtained by interpreting the inequality check (3.15) with the use of the H^1 norm:

$$|L(v)|^2 \leq \left[\int_0^1 |f(x)v(x)|dx\right]^2 \leq \|f\|_{L^2}^2 \|v\|_{L^2}^2 \leq \|f\|_{L^2}^2 \|v\|_{H^1}^2 . \tag{3.19}$$

Hence, the continuity constant that arises for the linear form L is equal to the L^2-norm of the second member f.

Finally, the property of the V-ellipticity for the bilinear form $a(.,.)$ is immediate if it is observed that:

$$a(v,v) = \|v'\|_{L^2}^2 + \|v\|_{L^2}^2 = \|v\|_{H^1}^2 \geq \|v\|_{H^1}^2 , \tag{3.20}$$

which means that the ellipticity constant is also equal to one.

The Lax-Milgram theorem then implies that there is strictly only one function belonging to $H_0^1(0,1)$, which is the solution of the variational problem (**VP**).

▶ *Remark*

The application of the Lax-Milgram could have been perfectly performed by choosing a different norm. Specifically, for this variational problem (**VP**) established in $H_0^1(0,1)$, a more precise norm exists to describe the elements of this Sobolev space.

Indeed, if the norm is changed by establishing:

$$\forall v \in H_0^1(0,1): \|v\|_{H_0^1} \equiv \left[\int_0^1 v'(x)^2 dx\right]^{1/2} . \tag{3.21}$$

It is then easily shown that (3.21) is a norm for $H_0^1(0,1)$.

In particular, the first property of the norms is fulfilled, given that any function v belonging to $H_0^1(0,1)$ is zero on the border of its definition interval.

Hence, if v is a function of $H_0^1(0,1)$ such that $\|v\|_{H_0^1} = 0$, v' is therefore zero on $[0,1]$. Consequently, v is a constant on the entire interval $[0,1]$ that can be measured, especially when $x = 0$. This then implies that v is identically zero on $[0,1]$.

It will be noted that the other properties of the norm $H_0^1(0,1)$ may be immediately established.

A.3) To achieve results of complementary regularity for weak solutions to a variational problem (**VP**) is often difficult and therefore requires the availability of a fairly sophisticated mathematical tool.

For the present work, the study is limited to simple one-dimensional case for which the mathematical tools that will be used are referred to in Chap. 1, paragraph 1.1, sect. 1.1.4.

Thus, let u be an element of $H_0^1(0,1)$, solution to the variational problem (**VP**) and the following is obtained:

$$\int_0^1 u'v'dx = \int_0^1 (f-u)v\,dx, \quad \forall v \in H_0^1(0,1).$$ (3.22)

The function space $C_0^1(]0,1[)$ is then introduced and defined by:

$$C_0^1(]0,1[) \equiv \{v: [0,1] \to \mathbf{R}, \; v \in C^1(]0,1[), \; Supp\, v \subset]0,1[\},$$ (3.23)

where $Supp\, v$ denotes the support of function v.

Therefore function v from $C_0^1(]0,1[)$ can be chosen in the equality (3.22), since this equality is in fact, a series of variational equations valid for all functions v belonging to $H_0^1(0,1)$ and containing $C_0^1(]0,1[)$.

Moreover, if function g is introduced and defined by $g = f - u$, the result is:

$$g \in L^2(0,1) \subset L^1(0,1) \subset L_{loc}^1(0,1),$$ (3.24)

where space $L_{loc}^1(0,1)$ is defined by:

Let any K be a closed subset strictly included in $[0,1]$, then:

$$\text{Given} \quad v \in L_{loc}^1(0,1) \quad \text{then} \quad v \in L^1(K).$$ (3.25)

Thus, the variational equations family (3.22) can be expressed in $C_0^1(]0,1[)$, (according to lemma 1), in the form of:

$$\int_0^1 u'v'dx = \int_0^1 gv\,dx = -\int_0^1 Gv'dx, \quad \forall v \in C_0^1(]0,1[),$$ (3.26)

where G is a primitive of g.

It is then expressed (3.26) in the form:

$$\int_0^1 (u'+G)v'dx = 0, \quad \forall v \in C_0^1(]0,1[).$$ (3.27)

Having finally obtained $u'+G \in L^2(0,1)$, since u belongs to $H^1(0,1)$, on one part, and G belongs to $C^0(]0,1[) \cap H^1(0,1)$ on the other, the result obtained, according to the same lemma 2, is $u'+G$ which belongs to $L_{loc}^1(0,1)$.

It is then possible to apply the lemma 1 to the variational equation (3.27), which leads to:

$$u'+G = C^{te}.$$ (3.28)

Thus, it is seen that u' is a function of $H^1(0,1)$, as the difference of a constant and of the G function.

It is finally inferred that u belongs to $H^2(0,1)$.

A.4) It is now important to establish the equivalence between the solution of the continuous problem **(CP)** and that of the variational problem **(VP)**.

Clearly, if u is a solution to the continuous problem **(CP)** looked for in $H^2(0,1)$ then u is a weak solution to the variational problem **(VP)**.

For this to happen, it is necessary that the construction process of the variational formulation **(VP)** be re-examined and be stated to be licit, (specially using the integration formula by parts, cf. Chapter 1, theorem 4), being given that u belongs to $H^2(0,1)$ and v to $H_0^1(0,1)$.

The reciprocal is then established. Let u be the solution belonging to $H_0^1(0,1) \cap H^2(0,1)$ of the variational problem **(VP)**.

The integration formula is used by parts in the reverse order to the one used to obtain the variational formulation.

This results in:

$$\int_0^1 (-u'' + u - f)v\,dx = 0, \quad \forall v \in H_0^1(0,1) . \tag{3.29}$$

The particular functions v belonging to $\mathscr{D}(0,1)$, which are the functions v belonging to $C^\infty(]0,1[)$ whose support is strictly included in interval $]0,1[$, are then chosen from the variational equations (3.29). This choice is legitimate since $\mathscr{D}(0,1) \subset H_0^1(0,1)$.

The density theorem 2 is then used as follows, rewriting equ. (3.29) in space $\mathscr{D}(0,1)$:

$$\int_0^1 (-u'' + u - f)v\,dx = 0, \quad \forall v \in \mathscr{D}(0,1) . \tag{3.30}$$

It would be better to have the last equality family (3.30) for any v function belonging to $L^2(0,1)$ in order to choose $v = -u'' + u - f$, as special function of $L^2(0,1)$ to conclude.

Therefore let φ be any function belonging to $L^2(0,1)$ and according to the density theorem 2, $\mathscr{D}(0,1)$ is dense in $L^2(0,1)$. Then, there exists a sequence of functions φ_n belonging to $\mathscr{D}(0,1)$ and that converge towards φ, according to the L^2 norm:

$$\lim_{n\to\infty} \left[\int_0^1 |\varphi_n - \varphi|^2\,dx \right] = 0 . \tag{3.31}$$

However, for any function of the sequence φ_n belonging to $\mathscr{D}(0,1)$, the equality (3.30) occurs by choosing $v = \varphi_n$:

$$\int_0^1 (-u'' + u - f)\varphi_n\,dx = 0, \quad \forall n \in \mathbf{N} . \tag{3.32}$$

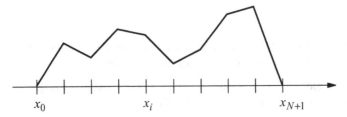

Fig. 3.1 Profile of a Piecewise Affine Function

It is then possible to obtain the same property for the functions φ of $L^2(0,1)$ by the following method:

$$\left| \int_0^1 \psi \cdot \varphi \right| = \left| \int_0^1 \psi \cdot (\varphi - \varphi_n) \right| \le \left[\int_\Omega |\psi|^2 \right]^{1/2} \cdot \left[\int_0^1 |\varphi_n - \varphi|^2 \right]^{1/2}, \qquad (3.33)$$

where it was stated: $\psi = -u'' + u - f$.

n is made to tend towards $+\infty$ in inequality (3.33), which demonstrates that:

$$\int_0^1 \psi \cdot \varphi \, dx = 0, \quad \forall \varphi \in L^2(0,1) . \qquad (3.34)$$

The demonstration now ends by choosing function as one of all functions of $L^2(0,1)$ which is equal to: $\varphi^* = -u'' + u - f$.

It is then deduced that:

$$-u'' + u + f = 0 \quad \text{in} \quad L^2(0,1) . \qquad (3.35)$$

Moreover, if the second member f belongs to $L^2(0,1) \cap C^0(]0,1[)$, then the differential equation is satisfied for any x belonging to $]0,1[$ and the solution u is the classical solution to the continuous problem **(CP)** belonging to $C^2(]0,1[)$.

▶ **Numerical part – Lagrange Finite Elements P_1**

A.5) The dimension of the approximation space $\tilde{V}$ can be determined in various ways. The simplest and smartest way is to state that the functions $\tilde{v}$ of $\tilde{V}$ are basically pecked lines affined by full mesh $[x_i, x_{i+1}]$ and cancelling each one when $x = 0$ and when $x = 1$, (see Fig. 3.1).

Hence, having $(N + 2)$ points of discretisation for the entire mesh of interval $[0,1]$, two $\tilde{V}$ functions stand out because of their difference in values that may be seen at N interior points $(x_1, \ldots, x_N)$.

Any function $\tilde{v}$ of $\tilde{V}$ also needs to satisfy, $\tilde{v}_0 = \tilde{v}_{N+1} = 0$.

In other words, a function $\tilde{v}$ belonging to $\tilde{V}$ is entirely determined by the N-tuple $(\tilde{v}_1, \ldots, \tilde{v}_N)$.

This implies that the space is isomorphic to $\mathbf{R}^N$. In conclusion, it can be deduced that the dimension of $\tilde{V}$ is equal to N.

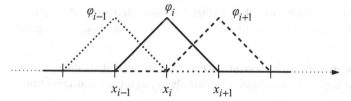

Fig. 3.2 Basis Functions φ_{i-1}, φ_i and φ_{i+1}

A.6) The approximated variational formulation is obtained by substituting the approximation functions $(\tilde{u}, \tilde{v})$ to the (u, v) functions in the variational formulation (**VP**).

Moreover, the approximation expressions given by (3.6) are used and the following is obtained:

Find the numerical sequence $(\tilde{u}_j)$, $(j = 1$ to $N)$, solution to:

$$\sum_{j=1,N} \left[\int_0^1 \left(\varphi_i' \varphi_j' + \varphi_i \varphi_j \right) dx \right] \tilde{u}_j = \int_0^1 f \varphi_i \, dx, \quad \forall i = 1 \text{ to } N. \tag{3.36}$$

The expressions of A_{ij}, and b_j corresponding to the formulas (3.8) are then obtained by identification.

▶ **Function φ_i Characteristic of a Node Strictly Interior at [0,1]**

A.7) The basis functions φ_i, characteristic of nodes strictly interior to integration interval $[0, 1]$, are now considered.

The generic equation of system (3.36) has, *a priori*, non zero terms, except those corresponding to φ_j functions whose support intercepts those of the φ_i function considered (see Fig. 3.2).

Thus, the basis functions concerned are: φ_{i-1}, φ_i and φ_{i+1}.

This explains why the equation ($\widetilde{\textbf{VP}_{\textbf{int}}}$), only has terms $A_{i,i-1}$, $A_{i,i}$ and $A_{i,i+1}$ and is expressed according to (3.9).

▶ **Approximate Calculation of Coefficients $A_{ij}, j = i - 1, i, i + 1$.**

a) *Approximation of coefficient A_{ii}.*

$$A_{ii} = \int_0^1 \left(\varphi_i'^2 + \varphi_i^2 \right) dx \quad = \int_{Supp \, \varphi_i} \left(\varphi_i'^2 + \varphi_i^2 \right) dx,$$

$$= \int_{x_{i-1}}^{x_i} \left(\varphi_i'^2 + \varphi_i^2 \right) dx \quad + \int_{x_i}^{x_{i+1}} \left(\varphi_i'^2 + \varphi_i^2 \right) dx,$$

$$\simeq \left(\frac{1}{h^2} \times h \right) + \frac{h}{2} (0+1) + \left(\frac{1}{h^2} \times h \right) + \frac{h}{2} (1+0), \tag{3.37}$$

$$A_{ii} \simeq \frac{2}{h} + h.$$

This was achieved by considering the fact that the basis functions φ_i of are piecewise affines. Thereafter, the derivatives φ_i' are constant on each mesh having the form $[x_i, x_{i+1}]$.

The integrals bearing on those derivatives can then be calculated either exactly or by using the trapezium quadrature formula being exact for constants functions.

b) *Approximation of coefficient $A_{i,i-1}$.*

$$A_{i,i-1} = \int_0^1 \left(\varphi_i' \varphi_{i-1}' + \varphi_i \varphi_{i-1} \right) dx ,$$

$$= \int_{Supp\, \varphi_{i-1} \cap Supp\, \varphi_i} \left(\varphi_i' \varphi_{i-1}' + \varphi_i \varphi_{i-1} \right) dx ,$$

$$\simeq \left(-\frac{1}{h^2} \times h \right) + \frac{h}{2} \left[(0 \times 1) + (1 \times 0) \right] , \tag{3.38}$$

$$A_{i,i-1} \simeq -\frac{1}{h} .$$

c) *Approximation of coefficient $A_{i,i+1}$.*

Calculation of the coefficient $A_{i,i+1}$ is easily obtained as long as the following symmetrical properties are observed:

– Matrix A of coefficient A_{ij} is symmetrical: $A_{i,j} = A_{j,i}$.

– The mesh over interval $[0, L]$ is translation invariant as a consequence of its uniform step of constant discretisation h.

It then becomes:

Symmetry   Invariant

$$\underset{\downarrow}{ } \qquad \underset{\downarrow}{ }$$

$$A_{i,i-1} \;\; = \;\; A_{i-1,i} \;\; = \;\; A_{i,i+1} \;\; \simeq \;\; -\frac{1}{h} .$$

▶ **Estimation of the Second Member b_i**

The second member b_i is calculated by considering that every basis function φ_i, characteristic of a strictly interior node has a support consisting of the union of the $[x_{i-1}, x_i]$ and $[x_i, x_{i+1}]$ intervals, (see Fig. 3.2).

It then becomes:

$$b_i = \int_0^1 f \varphi_i \, dx = \int_{x_{i-1}}^{x_i} f \varphi_i \, dx + \int_{x_i}^{x_{i+1}} f \varphi_i \, dx ,$$

$$\simeq \frac{h}{2} [0 + f_i] + \frac{h}{2} [f_i + 0] , \tag{3.40}$$

$$b_i \simeq h f_i .$$

A.8) The previous results (3.37)–(3.40) are then grouped to obtain the corresponding nodal equation:

$$-\frac{\tilde{u}_{i-1} - 2\tilde{u}_i + \tilde{u}_{i+1}}{h^2} + \tilde{u}_i = f_i, \quad (i = 1 \text{ to } N) . \tag{3.41}$$

A.9) Discretization by finite differences of second order differential equation of the continuous problem (**CP**) is classical.

The method uses the Taylor's formula after choosing to express the differential equation at the discretisation point x_i:

$$-u''(x_i) + u(x_i) = f(x_i), (i = 1 \text{ to } N) . \tag{3.42}$$

Taylor's formula enables the substitution of the second derivative u'' at point x_i by algebraic combination, using different values of the unknown u in different proximal points of the mesh.

To obtain an order which is consistent with the finite elements method, the progressive form and the regressive form of the Taylor's formula are used:

$$u(x_{i+1}) = u(x_i) + hu'(x_i) + \frac{h^2}{2}u''(x_i) + \frac{h^3}{6}u'''(x_i) + O(h^4) , \tag{3.43}$$

$$u(x_{i-1}) = u(x_i) - hu'(x_i) + \frac{h^2}{2}u''(x_i) - \frac{h^3}{6}u'''(x_i) + O(h^4) . \tag{3.44}$$

The addition of equs. (3.43) and (3.44) is then performed.

It then becomes:

$$u''(x_i) = \frac{u(x_{i+1}) - 2u(x_i) + u(x_{i-1})}{h^2} + O(h^2) . \tag{3.45}$$

The expression of the second derivative of u (3.45) is substituted at point x_i in the differential equation (3.20) to obtain:

$$-\frac{u(x_{i+1}) - 2u(x_i) + u(x_{i-1})}{h^2} + u(x_i) = f(x_i) + O(h^2), \quad (i = 1 \text{ to } N) . \tag{3.46}$$

The traces u_i of u, ($u_i \equiv u(x_i)$) are replaced at nodes x_i, by the approximations $\tilde{u}_i, (\tilde{u}_i \approx u_i)$, in order to preserve the equality between both members of (3.46) when suppressing the infinitely small $O(h^2)$.

This substitution leads to an exact correspondence between the scheme with finite differences and the nodal equation (3.41).

It is obvious that the scheme with finite differences (3.41) is of the second order, considering the approximation process that was explained earlier.

As a matter of fact, if the $u(x_i)$ values were substituted by the $\tilde{u}_i$ values, the result would be the differential equation (3.42) as closely as possible by $O(h^2)$.

This is the reason why the scheme having finite differences (3.41) is the second order.

3.2 The Neumann Problem

3.2.1 Statement

The aim of this problem is to propose a mathematical and numerical study of the solution to a linear differential problem, subjected to Neumann boundary conditions.

Thus, let u be a function of the real variable defined on $[0,1]$ and has values in $\mathbf{R}$. The interest is on the solution to the continuous problem (**CP**) defined by:

Find $u \in H^2(0,1)$ as solution to:

$$(\mathbf{CP}) \begin{cases} -u''(x) + u(x) = f(x), & 0 \le x \le 1, \\ u'(0) = u'(1) = 0, \end{cases} \qquad (3.47)$$

where f is a given function belonging to $L^2(0,1)$.

▶ Variational Formulation – Theoretical Part

1) Let v be a test function defined on $[0,1]$ having real values and belonging to a variational space V.

Show that the continuous problem (**CP**) can be written in a variational formulation (**VP**) like the following:

$$a(u,v) = L(v), \quad \forall v \in V.$$

The bilinear form $a(.,.)$, the linear form $L(.)$ and the functional space V will be specified.

2) Establish the existence and uniqueness of the weak solution of the variational problem (**VP**) in $H^1(0,1)$.

3) Show that any weak solution to the variational problem (**VP**) also belongs to $H^2(0,1)$.

4) Deduce from it the equivalence between the strong formulation of the problem (**CP**) set in $H^2(0,1)$ and the weak formulation of the variational problem (**VP**) considered in $H^1(0,1) \cap H^2(0,1)$.

▶ Numerical Part – Lagrange Finite Elements P_1

5) The approximation of the variational problem (**VP**) is performed by using Lagrange finite elements P_1.

To make that happen, we introduce a regular mesh of the interval $[0,1]$ with a constant step h, so that:

$$\begin{cases} x_0 = 0, \ x_{N+1} = 1 \, , \\ x_{i+1} = x_i + h, \ i = 0 \text{ to } N \, . \end{cases} \tag{3.48}$$

The approximation space $\tilde{V}$ is now defined using:

$$\tilde{V} = \left\{ \tilde{v} \colon [0,1] \to \mathbf{R}, \ \tilde{v} \in C^0([0,1]), \ \tilde{v}|_{[x_i, x_{i+1}]} \in P_1 \right\} \, , \tag{3.49}$$

where $P_1 \equiv P_1([x_i, x_{i+1}])$ refers to the space of polynomials defined over $[x_i, x_{i+1}]$ having a degree less than or equal to one.

– What is the dimension of $\tilde{V}$?

6) Let $\varphi_i, (i = 1 \text{ to } \dim \tilde{V})$, be the canonical basis of $\tilde{V}$ verifying $\varphi_i(x_j) = \delta_{ij}$, where δ_{ij} refers to the Kronecker symbol.

After having written the approximate variational formulation $(\widetilde{\textbf{VP}})$, having a solution and associated with the variational problem $(\textbf{VP})$, show that by selecting:

$$\tilde{v}(x) = \varphi_i(x), \quad (i = 1 \text{ to } \dim \tilde{V}) \quad \text{and} \quad \tilde{u}(x) = \sum_{j=1, \dim \tilde{V}} \tilde{u}_j \varphi_j \, , \tag{3.50}$$

the following $(\widetilde{\textbf{VP}})$ system is obtained:

$$(\widetilde{\textbf{VP}}) \ \sum_{j=1, \dim \tilde{V}} A_{ij} \tilde{u}_j = b_i, \quad \forall i \in \{1, \ldots, \dim \tilde{V}\} \, , \tag{3.51}$$

where the following was noted:

$$A_{ij} = \int_0^1 (\varphi_i' \varphi_j' + \varphi_i \varphi_j) \mathrm{d}x, \quad b_i = \int_0^1 f \varphi_i \mathrm{d}x \, . \tag{3.52}$$

▶ **Function φ_i Characteristic of Node Strictly Interior at [0,1]**

7) Given the mesh regularity, the generic nodal equation of the $(\widetilde{\textbf{VP}})$ system associated with any basis function $\varphi_i, (i = 1 \text{ to } \dim \tilde{V} - 2)$, characteristic of a node interior at $[0,1]$ is written as:

$$(\widetilde{\textbf{VP}}_{\textbf{Int}}) \ A_{i,i-1} \tilde{u}_{i-1} + A_{i,i} \tilde{u}_i + A_{i,i+1} \tilde{u}_{i+1} = b_i, \quad (\forall i = 1 \text{ to } \dim \tilde{V} - 2) \, . \tag{3.53}$$

– Using the trapezium formula, calculate the 4 coefficients $(A_{ij}, \ b_i)$.

8) Group the results together by writing the corresponding nodal equation.

9) Show that the centred finite differences scheme associated with the differential equation of the continuous problem **(CP)** is obtained again. What is its order of precision?

Remember that the trapezium quadrature is written as:

$$\int_a^b \xi(s)ds \simeq \frac{(b-a)}{2} \{\xi(a) + \xi(b)\} .$$

▶ **Function φ_0 Characteristic of the Node $x_0 = 0$**

10) When considering the basis function φ_0 characteristic of the node $x_0 = 0$, show that the corresponding nodal equation of the $(\widetilde{\mathbf{VP}})$ system is written as:

$$(\widetilde{\mathbf{VP_0}}) \; A_{0,0}\tilde{u}_0 + A_{0,1}\tilde{u}_1 = b_0 . \tag{3.54}$$

– Using the trapezium formula, calculate the 3 coefficients $A_{0,0}$, $A_{0,1}$ and b_0.

11) Write the corresponding nodal equation.

12) Show that the finite differences scheme associated with the Neumann condition in $x = 0$ is obtained again. What is its order of precision?

▶ **Function φ_{N+1} Characteristic of the Node $x_{N+1} = 1$**

13) Now, consider the basis function φ_{N+1} characteristic of the node $x_{N+1} = 1$.

Show that the nodal equation associated with the $(\widetilde{\mathbf{VP}})$ system is written as:

$$(\widetilde{\mathbf{VP_{N+1}}}) \; A_{N+1,N}\tilde{u}_N + A_{N+1,N+1}\tilde{u}_{N+1} = b_{N+1} . \tag{3.55}$$

– Using the trapezium formula, calculate the 3 coefficients $A_{N+1,N+1}$, $A_{N+1,N}$ and b_{N+1}.

14) Write the corresponding nodal equation.

15) Show that the finite differences scheme associated with the Neumann condition in $x = 1$ is obtained again. What is its order of precision?

3.2.2 Solution

▶ **Variational Formulation – Theoretical Part**

A.1) Let v be a test function defined on $[0, 1]$ having real values and "sufficiently regular".

As already mentioned in the presentation of the Dirichlet problem (see paragraph [3.1]), the regularity of functions v will be specified a *posteriori* in order to give sense to the variational formulation, when the latter is established.

The differential equation of the continuous problem (**CP**) is multiplied by v then integrated over the interval $[0, 1]$.

$$-\int_0^1 u''v\,dx + \int_0^1 uv\,dx = \int_0^1 fv\,dx, \quad \forall v \in V. \tag{3.56}$$

An integration by parts then gives the following:

$$\int_0^1 u'v'\,dx + u'(0)v(0) - u'(1)v(1) + \int_0^1 uv\,dx = \int_0^1 fv\,dx, \quad \forall v \in V. \tag{3.57}$$

Here, the homogenous Neumann boundary conditions defined in the continuous problem (**CP**), $(u'(0) = u'(1) = 0)$, appear in the integral formulation (3.57).

As a result and by considering the above two boundary conditions, the following formulation is obtained:

Find u belonging to V being the solution to:

$$\int_0^1 (u'v' + uv)\,dx = \int_0^1 fv\,dx, \quad \forall v \in V. \tag{3.58}$$

At this stage, the formulation (3.58) is only formal since the various integrals appearing in it have no reason to be convergent.

It is then observed that this variational formulation is strictly analogous to the one obtained within the framework of the Dirichlet problem – see paragraph [3.1], (3.13) – except the boundary conditions that should no longer be imposed on the test functions v within the framework of the Neumann problem treated here.

That is the reason why, if the functional analysis presented in paragraph [3.1] is used, a sufficient condition guaranteeing the convergence of the integrals in the variational formulation (3.58) consists in defining the variational space V as follows:

$$V \equiv H^1(0, 1) \equiv \{v: [0, 1] \to \mathbf{R}, v \in L^2(0, 1), v' \in L^2(0, 1)\}. \tag{3.59}$$

Finally, the variational problem (**VP**) is written as:

$$
(\text{VP}) \begin{cases}
\text{Find } u \text{ belonging to } V \text{ solution of: } a(u,v) = L(v), \quad \forall v \in V, \text{ where:} \\[2mm]
a(u,v) \equiv \displaystyle\int_0^1 \left[u'(x)v'(x) + u(x)v(x) \right] dx, \\[4mm]
L(v) \equiv \displaystyle\int_0^1 f(x)v(x)dx, \\[4mm]
V \equiv H^1(0,1).
\end{cases}
$$

$$(3.60)$$

A.2) The existence and uniqueness of the variational problem (**VP**) (3.60) are demonstrated by applying the Lax-Milgram theorem (lemma 4).

To make that happen, choosing one norm to be defined on the functional space $H^1(0,1)$ represents one of the key points in the application of the Lax-Milgram Theorem.

Then the choice is to measure the dimension of functions v belonging to $H^1(0,1)$ by the natural norm defined by:

$$
\forall v \in H^1(0,1): \|v\|_{H^1}^2 \equiv \int_0^1 v(x)^2 dx + \int_0^1 v'(x)^2 dx \equiv \|v\|_{L^2}^2 + \|v'\|_{L^2}^2. \quad (3.61)
$$

The norm being selected, the process that consists in verifying the various hypotheses of the Lax-Milgram theorem is strictly similar to the one presented for the Dirichlet problem, (see paragraph [3.1]).

The following is a point-by-point summary of this verification:

- The space $H^1(0,1)$ is a Hilbert space for the norm (3.61); the inner product resulting from this norm coincides exactly with the bilinear norm $a(.,.)$ defined by (3.60).

- The bilinear form $a(.,.)$ is continuous on $H^1(0,1) \times H^1(0,1)$; the continuity constant being equal to one.

- The linear form $L(.)$ defined by (3.60) is continuous on $H^1(0,1)$; the continuity constant being equal to the L^2-norm of the second member f.

- The form $a(.,.)$ is H^1-elliptic and the ellipticity constant is equal to one.

The application of the Lax-Milgram theorem thus implies the existence of one and only one function u belonging to $H^1(0,1)$, the solution to the variational problem (**VP**) defined by (3.60).

A.3) Once again, the Dirichlet problem [3.1] will be used to treat this question.

In fact, as mentioned above, achieving complementary regularity results for weak solutions to the variational problem (**VP**) may require sufficiently sophisticated mathematical tools.

That is the reason why the present study, being limited to a one-dimensional case, refers to mathematical results mentioned in Chap. 1, paragraph 1.1, sect. 1.1.4.

So, let u be an element of $H^1(0,1)$, solution to the variational problem (**VP**), and the following is obtained:

$$\int_0^1 u'v'dx = \int_0^1 (f-u)v\,dx, \quad \forall v \in H^1(0,1). \tag{3.62}$$

Among the functions v belonging to $H^1(0,1)$, only those belonging to $C_0^1(]0,1[)$, $(C_0^1(]0,1[) \subset H^1(0,1))$, are then selected.

Moreover, by introducing the function g defined by $g = f - u$, the following is obtained:

$$g \in L^2(0,1) \subset L^1(0,1) \subset L^1_{loc}(0,1), \tag{3.63}$$

where the space $L^1_{loc}(0,1)$ is defined by:

Let any K be a closed subset strictly included in $[0,1]$, then:

$$\text{Given } v \in L^1_{loc}(0,1) \text{ then } v \in L^1(K). \tag{3.64}$$

Thus the family of variational equations (3.62) can be written within $C_0^1(]0,1[)$ in the following form:

$$\int_0^1 u'v'dx = \int_0^1 gv\,dx = -\int_0^1 Gv'dx, \quad \forall v \in C_0^1(]0,1[), \tag{3.65}$$

where G is a primitive of g. (To make that happen, the lemma 2 would be used).

(3.65) is then written in the following form:

$$\int_0^1 (u'+G)v'dx = 0, \quad \forall v \in C_0^1(]0,1[). \tag{3.66}$$

Having finally obtained $u'+G \in L^2(0,1)$, (since, on one hand u belongs to $H^1(0,1)$ and on the other hand G belongs to $C^0(]0,1[) \cap H^1(0,1))$, still according to lemma 2, $u'+G$ belonging to $L^1_{loc}(0,1)$ is obtained in the same manner.

Then, lemma 1 may be applied to the variational equation (3.66), leading to:

$$u'+G = C^{te}. \tag{3.67}$$

Thus, it happens that u' is a function belonging to $H^1(0,1)$ as the difference of a constant and of the function G.

Finally, it is inferred that u belongs to $H^2(0,1)$.

A.4) This last question of the theoretical part is dedicated to the equivalence between the solution to the continuous problem (**CP**) and the solution to the variational problem (**VP**).

The direct way is simple since if u is a solution to the continuous problem (**CP**) searched for in $H^2(0,1)$, then u is a weak solution to the variational problem (**VP**).

To do so, it is only necessary to revert back to the process that enabled the establishment of the variational formulation (**VP**) and to note that the latter makes sense, (in particular by using the integration-by-parts formula, see Chap. 1, theorem 4), given that u belongs to $H^2(0,1)$ and v to $H^1(0,1)$.

The reciprocal is now calculated. In the variational problem (**VP**), let u be a solution belonging to $H^2(0,1)$ – any solution u to the variational problem henceforth belongs to $H^2(0,1)$ according to the previous question.

The integration-by-parts formula is then used in the inverse direction to the one that enabled the variational formulation to be obtained and this gives:

$$\int_0^1 (-u'' + u - f)v\,dx = 0, \quad \forall v \in H^1(0,1). \tag{3.68}$$

It is then noticed that the formulation (3.68) is identical to the one considered in the Dirichlet problem [3.1] except the functional framework (H_0^1 in the Dirichlet problem and H^1 which is here considered in the Neumann problem).

It is then only necessary to observe the functional inclusion $H_0^1 \subset H^1$ even if it is trivial, in order to strictly apply the whole methodology that has been presented. (see Dirichlet problem [3.1], question No. **4**).

The essential points treated in the rest of the demonstration is thus pointed out:

- In the equality (3.68), functions v belonging to $\mathscr{D}(0.1)$ are selected since $\mathscr{D}(0,1) \subset H^1(0,1)$.

- The density theorem 2 is used: $\mathscr{D}(0,1)$ is dense in $L^2(0,1)$.

- Then, it is shown that the equality (3.68) no longer takes place in $H^1(0,1)$ but in a bigger space i.e. in $L^2(0,1)$.

- It is then possible to choose from all the v functions belonging to $L^2(0,1)$ and involved in the equation (3.68), the one that exactly equals: $v^* = -u'' + u - f$.

If in addition, the second member f belongs to $L^2(0,1) \cap C^0(]0,1[)$ then the differential equation is satisfied for any $x \in]0,1[$ and the solution u is the classical solution to the continuous problem (**CP**) belonging to $C^2(]0,1[)$.

▶ **Numerical Part – Lagrange Finite Elements P_1**

A.5) To calculate the dimension of space $\tilde{V}$, the following remark is necessary:

The definition (3.49) of the approximation space is almost similar to the one considered in the Dirichlet problem (see problem of [3.1], question No.5, (3.5)).

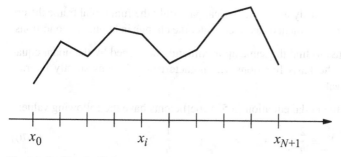

Fig. 3.3 Profile of a Piecewise Affine Function

Thus, using the demonstration performed within the framework of the Dirichlet problem, it is only necessary to note that in space $\tilde{V}$ defined by (3.49), two liberty degrees, due to the two values of any function $\tilde{v}$ of $\tilde{V}$ in $x = 0$ and in $x = 1$, add two units to the dimension found in the Dirichlet problem.

In other words, finding any function $\tilde{v}$ of $\tilde{V}$ means finding its trace $(\tilde{v}_0, \tilde{v}_1, \ldots, \tilde{v}_N, \tilde{v}_{N+1})$ in $(N+2)$ discretisation points of the mesh in the interval $[0, 1]$, i.e. $(x_0, x_1, \ldots, x_N, x_{N+1})$ which on their own fix the definition of $\tilde{v}$ (see Fig. 3.3).

As a result, space $\tilde{V}$ is isomorphic to $\mathbf{R}^{N+2}$ and the dimension of $\tilde{V}$ is equal to $N + 2$.

A.6) As usual, the approximate variational formulation $(\widetilde{\mathbf{VP}})$ is obtained by substituting the approximate functions $(\tilde{u}, \tilde{v})$ for functions (u, v) in the variational formulation $(\mathbf{VP})$.

Moreover, the expressions supplied by the formula (3.50) are used.

Thus, the approximate variational formulation $(\widetilde{\mathbf{VP}})$ is written as:

$$(\widetilde{\mathbf{VP}}) \left[\begin{array}{l} \text{Find the numerical sequence } (\tilde{u}_j), \ (j = 0 \text{ to } N+1), \text{ solution to:} \\ \displaystyle\sum_{j=0}^{N+1} \left[\int_0^1 \left(\varphi_i' \varphi_j' + \varphi_i \varphi_j \right) \right] \tilde{u}_j = \int_0^1 f \varphi_i(x), \quad \forall i = 0 \text{ to } N+1 . \end{array} \right. \quad (3.69)$$

The expressions of A_{ij}, and b_j corresponding to the formulas (3.52) are then obtained by identification.

▶ **Function φ_i characteristic of a node strictly interior at [0,1]**

A.7) When observing variational formulation $(\widetilde{\mathbf{VP}})$ defined by (3.69), it appears that, in the linear system consecutive equations $(N+2)$, the "interior" N equations corresponding to the values of j ranging from 1 to N, are totally identical to those found in the Dirichlet problem, (see paragraph 3.1, question **7**).

As mentioned previously in the theoretical part, only the functional frame differs between the two formulation in order to consider the change in boundary conditions.

It is then expected to find the same approximation described by the nodal equations associated to the basis functions φ_i, characteristic of nodes strictly interior to $[0, 1]$ mesh interval.

In other words, the nodal equation (3.53) coefficients have the following value:

$$A_{ii} \equiv \frac{2}{h} + h, \, A_{i,i-1} = A_{i,i+1} \equiv -\frac{1}{h}, \, b_i \equiv hf_i . \tag{3.70}$$

This is the result of the variational formulation identical formalism between the Dirichlet and the Neumann problem, for basis functions φ_i characteristic of nodes strictly interior at interval $[0, 1]$.

A.8) The nodal equation of the approximate variational problem $(\widetilde{\mathbf{VP}})$ corresponding to the basis function φ_i, which is characteristic of a strictly interior node x_i and written as:

$$-\frac{\tilde{u}_{i-1} - 2\tilde{u}_i + \tilde{u}_{i+1}}{h^2} + \tilde{u}_i = f_i, \, (i = 1 \text{ to } N) . \tag{3.71}$$

A.9) For the same reasons previously mentioned in the last questions, the analogy made with the Dirichlet problem ensures that the results, concerning the finite differences scheme obtained similarly in the present case, are at one's disposal.

Therefore, the finite differences scheme, with application of a second order discretisation to the differential equation of the continuous problem (**CP**), precisely corresponds to the nodal equation (3.71).

▶ **Function φ_0 Characteristic of the Node $x_0 = 0$**

A 10) The generic equation (3.69) of the approximate variational problem $(\widetilde{\mathbf{VP}})$, corresponding to the basis function φ_0, characteristic of the node x_0, is written as:

$$(\widetilde{\mathbf{VP_0}}) \, A_{00}\tilde{u}_0 + A_{01}\tilde{u}_1 = b_0 . \tag{3.72}$$

This results from the fact that when considering the approximate variational problem $(\widetilde{\mathbf{VP}})$ (3.69) in the case of the basis function φ_0, the summation upon the other basis functions φ_j only leads to zero contribution.

This is again the consequence of the position relative to the support of each of the basis functions φ_j as regards to the one of the basis function φ_0, (see Fig. 3.4).

▶ **Approximate Calculation of the Coefficients A_{00} and A_{01}**

a) *Approximation of the coefficient A_{00}.*

The calculation of the coefficient is performed in a way analogous to that presented for the calculation of the coefficient A_{ii} in the answer to the question **7**.

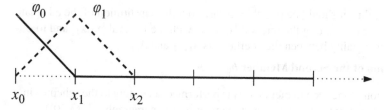

Fig. 3.4 Basis Functions φ_0 and φ_1

However, there is a difference since the basis function φ_0 comprises a support, which is solely constituted of the mesh $[x_0,x_1]$ while the basis functions φ_i, $(i = 1,N)$, have a support which is made up of two meshes, $[x_{i-1},x_i]$ and $[x_i,x_{i+1}]$.

The working out of the coefficient A_{00} is therefore performed as follows:

$$A_{00} = \int_0^1 \left(\varphi_0'^2 + \varphi_0^2 \right) dx = \int_{Supp\ \varphi_0} \left(\varphi_0'^2 + \varphi_0^2 \right) dx$$

$$= \int_{x_0}^{x_1} \left(\varphi_0'^2 + \varphi_0^2 \right) dx ,$$

$$\simeq \left(\frac{1}{h^2} \times h \right) + \frac{h}{2} \left(0 + 1 \right). \tag{3.73}$$

Thus finally:

$$A_{00} \simeq \frac{1}{h} + \frac{h}{2} . \tag{3.74}$$

b) *Approximation of the coefficient A_{01}.*

$$A_{01} = \int_0^1 \left(\varphi_0'\varphi_1' + \varphi_0\varphi_1 \right) dx = \int_{Supp\ \varphi_0 \cap Supp\varphi_1} \left(\varphi_0'\varphi_1' + \varphi_0\varphi_1 \right) dx ,$$

$$\simeq -\frac{1}{h^2} \times h + \frac{h}{2} \left[(0 \times 1) + (1 \times 0) \right]. \tag{3.75}$$

Results consequently in:

$$A_{01} \simeq -\frac{1}{h} . \tag{3.76}$$

It will be noted that the approximation of the coefficient A_{01} represents a particular case of the generic calculation presented above (see paragraph 3.1, (3.38), in so far as the discretisation step h is constant.

Therefore, the integration of the basis function φ_0 against the basis function φ_1 along the interval $[x_0,x_1]$ is completely equivalent to the integration of φ_i against φ_{i+1} along the interval $[x_i,x_{i+1}]$.

To prove this, it should be possible to carry out the substitution of the adequate variable by associating the interval $[x_i, x_{i+1}]$ with the interval $[x_0, x_1]$ and to observe the equality between the coefficients $A_{i,i+1}$ and A_{01}.

▶ **Evaluation of the Second Member b_0**

The evaluation of the second member b_0 is performed according to the scheme similar to the one presented for the second member b_i (see paragraph 3.1, (3.40)).

Therefore, the following is obtained:

$$b_0 = \int_0^1 f\varphi_0 \, dx = \int_{x_0}^{x_1} f\varphi_0 \, dx \simeq \frac{h}{2}[0 + f_0] \simeq \frac{h}{2}f_0 . \tag{3.77}$$

A.11) The results (3.73)–(3.77) are consequently gathered to obtain the corresponding nodal equation associated with the function of φ_0, characteristic of the node x_0.

$$\left(\frac{1}{h} + \frac{h}{2}\right)\tilde{u}_0 - \frac{1}{h}\tilde{u}_1 = \frac{h}{2}f_0 . \tag{3.78}$$

A.12) To determine the finite differences scheme which discretises the Neumann condition when $x_0 = 0, (u'(x_0) = 0)$, it can be observed that, if it were necessary to maintain the second order of discretisation obtained for the finite differences scheme (3.72), associated with differential equation of the continuous problem **(CP)**, inside the interval $[0, 1]$, the Taylor's expansion that will be studied must be written till the third order.

Therefore, the following Progressive Taylor's Expansion will be written as follows:

$$u(x_1) = u(x_0) + hu'(x_0) + \frac{h^2}{2}u''(x_0) + O(h^3) . \tag{3.79}$$

It is then possible to replace the value of the first derivative which is zero when $x_0 = 0$ (since it concerns the Neumann condition), still, as usual, in the case of the application of such a method, there appears the second derivative of u at the point x_0.

Therefore, the differential equation of the continuous problem **(CP)** can be assumedly written till the border of the interval, namely, here, when $x_0 = 0$:

$$-u''(x_0) + u(x_0) = f(x_0) . \tag{3.80}$$

The equ. (3.80) can express the second derivative u'' at the point x_0 and inserted in the Taylor's expansion (3.79).

There consequently results:

$$u(x_1) = u(x_0) + hu'(x_0) + \frac{h^2}{2}[u(x_0) - f(x_0)] + O(h^3) . \tag{3.81}$$

The approximations can then be worked out, while omitting the rest $O(h^3)$ in the equ. (3.81).

Following the process of discretisation when $x_0 = 0$, the equation is then written as:

$$\tilde{u}_1 = \tilde{u}_0 + \frac{h^2}{2} [\tilde{u}_0 - f_0] . \qquad (3.82)$$

The nodal equation (3.78), corresponding to the basis function φ_0, characteristic of the node x_0 of the discretisation is found.

▶ Function φ_{N+1} characteristic of the node $x_{N+1} = 1$

A.13) Considerations analogous to those presented above, for the working out of the nodal equation when $x_0 = 0$, are also relevant for the nodal equation when $x_{N+1} = 1$.

As such, it is only necessary to see to it that the situation of the basis functions φ_N and φ_{N+1} is symmetrical to the situation of the basis functions φ_0 and φ_1, (see Fig. 3.5).

That is why the equation of the approximate variational formulation $(\widetilde{VP})$ corresponding to the basis function φ_{N+1} is written as:

$$(\widetilde{VP}_{N+1}) \quad A_{N+1,N}\tilde{u}_N + A_{N+1,N+1}\tilde{u}_{N+1} = b_{N+1} . \qquad (3.83)$$

Likewise, provided that a great care is taken to replace the integration interval $[x_0, x_1]$ by the one which corresponds to the support of the function φ_{N+1} i.e. $[x_N, x_{N+1}]$, the following results for the evaluation of coefficients $A_{N+1,N}$, $A_{N+1,N+1}$ and b_{N+1} are obtained.

▶ Approximate Calculation of Coefficients $A_{N+1,N}$ and $A_{N+1,N+1}$

a) *Approximation of coefficient $A_{N+1,N+1}$.*

$$A_{N+1,N+1} = \int_0^1 \left(\varphi_{N+1}'^2 + \varphi_{N+1}^2 \right) dx = \int_{Supp\, \varphi_{N+1}} \left(\varphi_{N+1}'^2 + \varphi_{N+1}^2 \right) dx ,$$

$$= \int_{x_N}^{x_{N+1}} \left(\varphi_{N+1}'^2 + \varphi_{N+1}^2 \right) dx ,$$

$$\simeq \left(\frac{1}{h^2} \times h \right) + \frac{h}{2} (0 + 1) ,$$

$$\simeq \frac{1}{h} + \frac{h}{2} . \qquad (3.84)$$

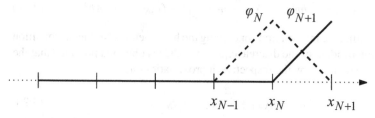

Fig. 3.5 Basis Functions φ_N and φ_{N+1}

b) *Approximation of coefficient $A_{N+1,N}$.*

$$A_{N+1,N} = \int_0^1 \left(\varphi_N' \varphi_{N+1}' + \varphi_N \varphi_{N+1} \right) dx ,$$

$$= \int_{Supp\ \varphi_N\ \cap\ Supp\ \varphi_{N+1}} \left(\varphi_N' \varphi_{N+1}' + \varphi_N \varphi_{N+1} \right) dx ,$$

$$\simeq \left(-\frac{1}{h^2} \times h \right) + \frac{h}{2} \left[(0 \times 1) + (1 \times 0) \right] \simeq -\frac{1}{h} . \tag{3.85}$$

▶ **Estimation of the Second Member b_{N+1}**

The approximation of the second member b_{N+1} is obtained by using the trapezium quadrature formula:

$$b_{N+1} = \int_0^1 f \varphi_{N+1} dx ,$$

$$= \int_{x_N}^{x_{N+1}} f \varphi_{N+1} dx \simeq \frac{h}{2} \left[0 + f_{N+1} \right] \simeq \frac{h}{2} f_{N+1} . \tag{3.86}$$

A.14) The nodal equation associated with the basis function φ_{N+1} is written by grouping the results of (3.84)–(3.86).

This equation is perfectly symmetrical compared to the one associated with the basis function φ_0:

$$-\frac{1}{h} \tilde{u}_N + \left(\frac{1}{h} + \frac{h}{2} \right) \tilde{u}_{N+1} = \frac{h}{2} f_{N+1} . \tag{3.87}$$

A.15) Given the symmetry mentioned in the previous question, finding the nodal equation (3.87) by using finite differences is conceivable.

In fact, at abscissa $x_{N+1} = 1$, if a progressive expansion was considered when $x_0 = 0$, this time a regressive Taylor's expansion must be considered in the following way:

$$u(x_N) = u(x_{N+1}) - hu'(x_{N+1}) + \frac{h^2}{2} u''(x_{N+1}) + O(h^3) . \tag{3.88}$$

Then, the second derivative when x_{N+1} is replaced by writing the differential equation of the continuous problem (**CP**) at the point x_{N+1}:

$$u(x_N) = u(x_{N+1}) - hu'(x_{N+1}) + \frac{h^2}{2} \left[u(x_{N+1}) - f(x_{N+1}) \right] + O(h^3) . \tag{3.89}$$

Now, by exploiting the information concerning the homogenous Neumann condition when x_{N+1}, the pendant of the discrete equation (3.82) is obtained provided that the values $u(x_i)$ are replaced by the respective approximations $\tilde{u}_i$:

$$\tilde{u}_N = \tilde{u}_{N+1} + \frac{h^2}{2} \left[\tilde{u}_{N+1} - f_{N+1} \right] . \tag{3.90}$$

3.3 The Fourier-Dirichlet Problem

3.3.1 Statement

The aim of this problem is to propose a mathematical and numerical study of the solution to a second order linear differential problem subjected to Fourier-Dirichlet mixed boundary conditions.

Let u be a function of a real variable, defined from values $[0,1]$ in $\mathbf{R}$.

The considered continuous problem (**CP**) is defined by:

To find $u \in H^2(0,1)$ which is the solution to:

$$(\mathbf{CP}) \begin{cases} -u''(x) + u(x) = f(x), & 0 \le x \le 1, \\ u(0) = 0, & u'(1) + ku(1) = 1, \end{cases} \qquad (3.91)$$

where f is a given function belonging to $L^2(0,1)$ and k a given positive or zero real parameter.

▶ **Variational Formulation – Theoretical Part**

1) Let v be a test function, defined by $[0,1]$, and having real values, belonging to the variational space V. Show that the continuous problem (**CP**) can be expressed as a variational formulation (**VP**) in the form:

$$a(u,v) = L(v), \quad \forall v \in V.$$

The bilinear form $a(.,.)$, the linear form $L(.)$ and the functional space V need to be specified.

2) Establish the existence and uniqueness of a weak solution of the variational problem (**VP**) in $H_*^1(0,1)$ defined by:

$$H_*^1(0,1) = \left\{ v :]0,1[\to \mathbf{R}, v \text{ and } v' \in L^2(0,1), v(0) = 0 \right\}. \qquad (3.92)$$

3) Show that any weak solution to $H_*^1(0,1)$ the variational problem (**VP**) also belongs to $H^2(0,1)$.

4) Infer from therein, the equivalence between the strong formulation presented in $H^2(0,1)$ and the weak formulation considered in $H_*^1(0,1) \cap H^2(0,1)$.

▶ **Lagrange Finite Element P_1 – Numerical Part**

5) Approximation of the variational problem (**VP**) is performed using Lagrange finite elements P_1.

To achieve this, a regular mesh of interval $[0,1]$ of constant step h is introduced, such as:

$$\begin{cases} x_0 = 0, \ x_{N+1} = 1 , \\ x_{i+1} = x_i + h, \ i = 0 \text{ to } N . \end{cases} \quad (3.93)$$

The approximation space $\tilde{V}$ is now defined using:

$$\tilde{V} = \left\{ \tilde{v} \colon [0,1] \rightarrow \mathbf{R}, \ \tilde{v} \in C^0([0,1]), \ \tilde{v}|_{[x_i,x_{i+1}]} \in P_1([x_i,x_{i+1}]), \ \tilde{v}(0) = 0 \right\} , \quad (3.94)$$

where $P_1([x_i,x_{i+1}])$ denotes the polynomial space defined over $[x_i,x_{i+1}]$, of degree less than or equal to one.

– What is the dimension of $\tilde{V}$?

6) Let $\varphi_i, (i = 1 \text{ to } \dim\tilde{V})$, be the canonical basis $\tilde{V}$ of establishing $\varphi_i(x_j) = \delta_{ij}$.

After having written the approximate variational formulation $(\widetilde{\mathbf{VP}})$, of solution $(\widetilde{\mathbf{VP}})$, which is associated to the variational problem $(\mathbf{VP})$, show that by choosing:

$$\tilde{v}(x) = \varphi_i(x), \ (i = 1 \text{ to } \dim\tilde{V}) \text{ and } \tilde{u}(x) = \sum_{j=1,\, \dim \tilde{V}} \tilde{u}_j \varphi_j , \quad (3.95)$$

the following $(\widetilde{\mathbf{VP}})$ system is obtained:

$$(\widetilde{\mathbf{VP}}) \quad \sum_{j=1,\, \dim \tilde{V}} A_{ij}\tilde{u}_j = b_i, \quad \forall i \in \{1,\dots, \dim \tilde{V}\} , \quad (3.96)$$

where the following was noted:

$$A_{ij} = \int_0^1 (\varphi_i' \varphi_j' + \varphi_i \varphi_j)\mathrm{d}x + k\varphi_i(1)\varphi_j(1), \quad b_i = \int_0^1 f\varphi_i \mathrm{d}x + \varphi_i(1) . \quad (3.97)$$

▶ **Function φ_i Characteristic of a Node Strictly Interior at [0,1]**

7) Given the regularity of the mesh, the generic nodal equation of the $(\widetilde{\mathbf{VP}})$ system associated to any function with basis φ_i, which is characteristic of a node strictly interior at $[0,1]$, is expressed as:

$$(\widetilde{\mathbf{VP}}_{\text{Int}}) \ A_{i,i-1}\tilde{u}_{i-1} + A_{i,i}\tilde{u}_i + A_{i,i+1}\tilde{u}_{i+1} = b_i,$$
$$(\forall i = 1, \dim \tilde{V} - 1) . \quad (3.98)$$

– Using the trapezium rule, calculate the 4 coefficients (A_{ij}, b_i).

8) Group the results together by writing down the corresponding nodal equation.

9) Show that the centred finite differences scheme associated with the differential equation of the continuous problem (**CP**) is obtained again. What is its order of precision? It is pointed out that the trapezium quadrature formula is written as:

$$\int_a^b \xi(s)ds \simeq \frac{(b-a)}{2}\{\xi(a)+\xi(b)\} .$$

▶ **Function φ_{N+1} Characteristic of the Node $x_{N+1} = 1$**

10) Now, consider the basis function φ_{N+1} characteristic of the node $x_{N+1} = 1$.

Show that the nodal equation associated with the ($\widetilde{\mathbf{VP}}$) system is written as:

$$(\widetilde{\mathbf{VP}}_{\mathbf{N+1}}) \quad A_{N+1,N}\tilde{u}_N + A_{N+1,N+1}\tilde{u}_{N+1} = b_{N+1} . \qquad (3.99)$$

- Using the trapezium formula, calculate the 3 coefficients $A_{N+1,N}$, $A_{N+1,N+1}$ and b_{N+1}.

11) Write the corresponding nodal equation.

12) Show that the finite differences scheme associated to the Fourier boundary conditions when $x = 1$ is obtained again. What is its order of precision?

3.3.2 Solution

▶ **Variational Formulation – Theoretical Part**

A.1) Let v be a test function defined by $[0,1]$ having real values and "sufficiently regular".

As already mentioned in the presentation of the Dirichlet problem (see paragraph [3.1]), the regularity of functions v will be specified *a posteriori* in order to give sense to the variational formulation, when the latter is established.

The differential equation of the continuous problem **(CP)** is multiplied by v then integrated over the interval $[0,1]$.

$$-\int_0^1 u''(x)v(x)\mathrm{d}x + \int_0^1 u(x)v(x)\mathrm{d}x = \int_0^1 f(x)v(x)\mathrm{d}x, \quad \forall v \in V. \tag{3.100}$$

The integration by parts then results in:

$$\int_0^1 u'(x)v'(x)\mathrm{d}x + u'(0)v(0) - u'(1)v(1) + \int_0^1 u(x)v(x)\mathrm{d}x,$$

$$= \int_0^1 f(x)v(x)\mathrm{d}x, \quad \forall v \in V. \tag{3.101}$$

Here the Fourier boundary conditions, defined in the continuous problem **(CP)**, $(u'(1) + ku(1) = 1)$, appears in the integral formulation (3.101). In fact, this may be confirmed by re-writing the Fourier condition in the form:

$$u'(1) = 1 - ku(1), \tag{3.102}$$

in order to replace the first derivative of solution u, at the abscissa $x = 1$ in equation (3.101).

Moreover, as previously observed for this type of second order differential equation, the homogenous Dirichlet condition when $x = 0$ cannot be directly taken into account in formulation (3.101).

This explains why this homogenous Dirichlet condition is imposed on test functions v whose solution u constitutes a specific case.

This method guarantees that the full memory of the information contained in the continuous problem **(CP)** is maintained in the future variational formulation.

These two boundary conditions, (the first one bearing on u via its relationship with (3.102) and the second one concerning the zero v test functions when $x = 0$) lead to the following variational formulation **(VP)**:

Find u belonging to V, solution to:

$$\int_0^1 (u'v' + uv)\mathrm{d}x + ku(1)v(1) = \int_0^1 fv\mathrm{d}x + v(1), \quad \forall v \in V. \tag{3.103}$$

At this stage of the study, the V space is composed of v functions subjugated to the homogenous Dirichlet condition when $x = 0 : v(0) = 0$.

Meanwhile, formulation (3.103) is only formal since the different integrals do not need to be convergent.

Actually, this variational formulation is structurally analogous to the one obtained within the framework of the Dirichlet problem – see (3.13), paragraph [3.1] – except for the boundary conditions that need to be modified to suit the test functions v within the framework of the Fourier-Dirichlet problem being examined here.

Thus, referring to the functional analysis previously shown in paragraph [3.1], a sufficient condition securing the convergence of integrals of the variational formulation (3.103) is to consider the following functional framework:

$$V \equiv H^1(0,1) \equiv \{v \colon [0,1] \to \mathbf{R}, v \in L^2(0,1), v' \in L^2(0,1)\}. \tag{3.104}$$

The homogenous Dirichlet condition at abscissa $x = 0$ is added to the above functional space to finally lead to the following variational formulation:

$$(\mathbf{VP}) \begin{cases} \text{Find } u \text{ belonging to } V \text{ solution of: } a(u,v) = L(v), \quad \forall v \in V, \text{ where:} \\[2mm] a(u,v) \equiv \displaystyle\int_0^1 (u'v' + uv)\mathrm{d}x + ku(1)v(1), \\[2mm] L(v) \equiv \displaystyle\int_0^1 fv\mathrm{d}x + v(1), \\[2mm] V \equiv H_*^1(0,1), \end{cases}$$

$$\tag{3.105}$$

where space $H_*^1(0,1)$ is defined by (3.92).

A.2) The existence and uniqueness of the solution to the variational problem (**VP**) defined by (3.105), is obtained by applying the Lax-Milgram theorem 4.

This is achieved by again choosing the $H^1(0,1)$ norm (see paragraph 3.1, (3.61) and (3.106) defined hereafter), by trading off measurement of the "size" of as any function of $H^1(0,1), (H_*^1(0,1) \subset H^1(0,1))$.

The $H^1(0,1)$ natural norm, previously defined, is as follows:

$$\forall v \in H^1(0,1) \colon \|v\|_{H^1}^2 \equiv \int_0^1 v(x)^2\mathrm{d}x + \int_0^1 v'(x)^2\mathrm{d}x \equiv \|v\|_{L^2}^2 + \|v'\|_{L^2}^2. \tag{3.106}$$

Once the norm is chosen, the different points described below need to be validated in order to apply the Lax-Milgram theorem:

a) $H_*^1(0,1)$ space is a Hilbert space for norm (3.106).

This is achieved by showing, for example, that $H_*^1(0,1)$ is a closed vector subspace of $H^1(0,1)$ for norm (3.106), (see H. Brézis, [1]).

b) The bilinear form $a(.,.)$ is continuous on $H^1_*(0,1) \times H^1_*(0,1)$ for norm (3.106).

In fact, let be $(u,v) \in H^1_*(0,1) \times H^1_*(0,1)$:

$$
\begin{aligned}
|a(u,v)| &\leq |(u,v)_{H^1}| + k|u(1)v(1)| \\
&\leq (u,u)^{1/2}_{H^1} \cdot (v,v)^{1/2}_{H^1} + k|u(1)v(1)| \\
&\leq \|u\|_{H^1} \cdot \|v\|_{H^1} + k|u(1)v(1)|,
\end{aligned}
\tag{3.107}
$$

where $(.,.)_{H^1}$ denotes the inner product of H^1 from which norm (3.106) arises.

Moreover, for any function v belonging to $H^1(0,1)$, the result is:

$$
\forall x \in [0,1]: v(x) = v(0) + \int_0^x v'(t)dt .
\tag{3.108}
$$

Thus, for the specific case of v functions belonging to $H^1_*(0,1)$, $v(0) = 0$.

It then becomes:

$$
\forall x \in [0,1]: v(x) = \int_0^x v'(t)dt .
\tag{3.109}
$$

Equation (3.109) is then expressed as $x = 1$ and the Cauchy-Schwartz inequality is used to obtain the following control inequality:

$$
|v(1)| \leq \left[\int_0^1 v'^2(t)dt \right]^{1/2} \equiv \|v'\|_{L^2} .
\tag{3.110}
$$

Inequality (3.107) is then used to obtain:

$$
|a(u,v)| \leq \|u\|_{H^1} \cdot \|v\|_{H^1} + k\|u'\|_{L^2}\|v'\|_{L^2} \leq (1+k)\|u\|_{H^1} \cdot \|v\|_{H^1} .
\tag{3.111}
$$

The bilinear form $a(.,.)$ is thus continuous for the H^1-norm defined by (3.106). The continuity constant is actually equal to $(1+k)$.

c) The linear form $L(.)$ defined by (3.105) is continuous on $H^1_*(0,1)$ for norm (3.106).

The same scheme of analysis previously presented for the bilinear form $a(.,.)$ is applied.

$$
|L(v)| \leq \int_0^1 |fv|dx + |v(1)| \leq \|f\|_{L^2}\|v\|_{L^2} + |v(1)| .
\tag{3.112}
$$

The inequality control (3.110), which is valid for any v function belonging to $H^1_*(0,1)$, is used again to obtain:

$$
|L(v)| \leq \|f\|_{L^2}\|v\|_{L^2} + \|v'\|_{L^2} \leq (1+\|f\|_{L^2})\|v\|_{H^1} .
\tag{3.113}
$$

The linear form $L(.)$ is therefore continuous on space $H^1_*(0,1)$ with the norm (3.106) and the continuity constant is equal to $(1+\|f\|_{L^2})$.

d) The $a(.,.)$ form is H^1-elliptic and the ellipticity constant is equal to one.

In fact, the ellipticity inequality is immediate if it is reckoned that:

$$a(v,v) = \|v'\|^2_{L^2} + \|v\|^2_{L^2} + kv(1)^2 = \|v\|^2_{H^1} + kv(1)^2 \geq \|v\|^2_{H^1} . \tag{3.114}$$

The fact that parameter k is a real positive number has been used.

The application of the Lax-Milgram theorem (see theorem 10) thus implies that there exists one and only one function belonging to $H^1_*(0,1)$ being the solution to the variational problem **(VP)** defined by (3.105).

A.3) The method presented for the Dirichlet problem [3.1] is used to prove that any weak solution belonging to $H^1_*(0,1)$ is also a function of $H^2(0,1)$.

Thus, let u be an element of $H^1_*(0,1)$, solution to the variational problem **(VP)** and the following is obtained:

$$\int_0^1 u'v'\mathrm{d}x = \int_0^1 (f-u)v\mathrm{d}x + [1-ku(1)]v(1), \quad \forall v \in H^1_*(0,1) . \tag{3.115}$$

Among the functions v belonging to $H^1_*(0,1)$, those belonging to $C^1_0(]0,1[)$, $(C^1_0(]0,1[) \subset H^1_*(0,1))$, are then selected.

This choice implies that the retained functions v are null when $x=0$ and when $x=1$.

Moreover, the function g defined by $g=f-u$ is introduced and the following is obtained:

$$g \in L^2(0,1) \subset L^1(0,1) \subset L^1_{\mathrm{loc}}(0,1) , \tag{3.116}$$

where space $L^1_{\mathrm{loc}}(0,1)$ is defined by:

Let any K be a closed subset strictly included in $]0,1[$, then:

$$\text{Given} \quad v \in L^1_{\mathrm{loc}}(0,1) \quad \text{then} \quad v \in L^1(K) . \tag{3.117}$$

Thus, the family of variational equations (3.115) can be written within $C^1_0(]0,1[)$ in the form:

$$\int_0^1 u'(x)v'(x)\mathrm{d}x = \int_0^1 g(x)v(x)\mathrm{d}x = -\int_0^1 G(x)v'(x)\mathrm{d}x, \quad \forall v \in C^1_0(]0,1[) , \tag{3.118}$$

where G is a primitive of g (Lemma 2 is used to achieve this).

(3.118) is then written in the form:

$$\int_0^1 \left[u'(x) + G(x) \right] v'(x) \mathrm{d}x = 0, \quad \forall v \in C_0^1(0,1) . \tag{3.119}$$

In this form, the family of variational equations (3.119) strictly corresponds to the family of equations demonstrated in the Dirichlet problem (see Dirichlet problem [3.1], (3.27)).

The rest of the analysis is inferred from it and the same arguments are used to state that any solution u to the variational problem belonging to $H_*^1(0,1)$ also belongs to the Sobolev space $H^2(0,1)$.

A.4) This last question of the theoretical part is dedicated to the equivalence between the solution to the continuous problem (**CP**) and the solution to the variational problem (**VP**).

The direct sense is simple as it is the result of the construction of a solution to the variational formulation (**VP**) using a given solution to the continuous problem (**CP**).

Then, it will be observed that this construction process is licit provided that the solution u to the continuous problem (**CP**) belongs to $H^2(0,1)$ and the test functions v that intervene in the variational formulation (**VP**) belong to $H_*^1(0,1)$.

It will notably be noticed that if u is a solution to the continuous problem (**CP**) satisfying the homogenous Dirichlet boundary conditions when $x = 0$, then this time u when considered as a solution to the variational problem (**VP**) belongs *de facto* to $H_*^1(0,1)$.

The reciprocal is now considered. Let u belong to $H^2(0,1) \cap H_*^1(0,1)$ a solution to the variational problem (**VP**). According to the previous question, it is known that any solution u of the variational problem (**VP**) belonging to $H_*^1(0,1)$ also belongs to $H^2(0,1)$.

The integration-by-parts formula used in the reverse order to the one that yielded the variational formulation (**VP**) leads to the following:

$$\int_0^1 (-u'' + u - f)v \, \mathrm{d}x + \left[u'(1) + ku(1) - 1 \right] v(1) = 0, \quad \forall v \in H_*^1(0,1) . \tag{3.120}$$

Then, consider the particular case of functions v in the equation (3.120) belonging to $H_0^1(0,1)$. This means that the functions v that are null when $x = 0$ and when $x = 1$.

In that case, the equation (3.120) is written as:

$$\int_0^1 (-u'' + u - f)v \, \mathrm{d}x = 0, \quad \forall v \in H_0^1(0,1) . \tag{3.121}$$

Here it is observed that, the formulation (3.121) is similar to the one considered in the Dirichlet problem [3.1].

That is why the rest of the reasoning is also similar. The essential points for the continuation of the demonstration are:

- In the equality (3.121), functions v belonging to $\mathscr{D}(0,1)$ are selected since $\mathscr{D}(0,1) \subset H^1(0,1)$.
- The density theorem 2 is used: $\mathscr{D}(0,1)$ is dense in $L^2(0,1)$.
- Then, it is shown that the equality (3.121) no longer takes place in $H^1(0,1)$ but in a bigger space i.e. in $L^2(0,1)$.
- It is then possible to choose from all the v functions belonging to $L^2(0,1)$ the one that exactly equals: $v^* = -u'' + u - f$.

Moreover, if the second member f belongs to $L^2(0,1) \cap C^0(]0,1[)$, the differential equation is satisfied for any x belonging to $]0,1[$ and the solution u is the classical solution of the continuous problem (**CP**) belonging to $C^2(]0,1[)$.

Once it is proved that the differential equation of the problem (**CP**) is satisfied by the solution to the variational problem (**VP**), the family of equations (3.120) is reduced and is written as:

$$[u'(1) + ku(1) - 1] v(1) = 0, \quad \forall v \in H_*^1(0,1). \tag{3.122}$$

A last selection in equation (3.122) consists in considering the particular case of function v^* defined by: $v^{**}(x) = x, \forall x \in [0,1]$.

Of course, it will be easily verified that v^{**} belongs to $H_*^1(0,1)$. In that case, equation (3.122) implies that u satisfies the Fourier boundary conditions:

$$u'(1) + ku(1) = 1. \tag{3.123}$$

This ends the demonstration of the reciprocal and any solution of the variational problem (**VP**) belonging to $H^2(0,1) \cap H_*^1(0,1)$ is a solution to the continuous problem (**CP**).

► **Lagrange Finite Elements P_1 – Numerical Part**

A.5) To calculate the dimension of space $\tilde{V}$, the following observation is necessary:

Once again, definition (3.94) of the approximation space $\tilde{V}$ is very close to the one considered in the Dirichlet problem (see problem of [3.1], question **5**, (3.5)).

Thus, by considering the demonstration presented above, it is only necessary to observe that space $\tilde{V}$ defined by (3.94) produces an additional degree of freedom attributable to the value of function $\tilde{v}$ of $\tilde{V}$ when $x = 1$.

As a result, the dimension of $\tilde{V}$ defined by (3.5) found in the Dirichlet problem should be increased by one unit.

In other words, for any function $\tilde{v}$ of $\tilde{V}$, knowledge of its trace $(\tilde{v}_1, \ldots, \tilde{v}_{N+1})$ at $(N+1)$ discretisation points of the mesh at the $[0,1]$ interval, namely in $(x_1, \ldots, x_{N+1})$, that fixes the definition of $\tilde{v}$ in a unique manner.

That is why space $\tilde{V}$ is isomorphic to $\mathbf{R}^{N+1}$ and the dimension of $\tilde{V}$ is equal to $N+1$.

A.6) In order to obtain the approximate variational formulation $(\widetilde{\mathbf{VP}})$, the approximation functions $(\tilde{u}, \tilde{v})$ are substituted in the (u, v) functions of the variational formulation $(\mathbf{VP})$. Moreover, the expressions given by (3.95) are used.

The approximate variational formulation $(\widetilde{\mathbf{VP}})$ is thus written as:

Find the numerical sequence $(\tilde{u}_j), (j = 1 \text{ to } N+1)$, solution to:

$$\sum_{j=1}^{N+1} \left[\int_0^1 \left(\varphi_i' \varphi_j' + \varphi_i \varphi_j \right) dx + k\varphi_i(1)\varphi_j(1) \right] \tilde{u}_j = \int_0^1 f\varphi_i(x)dx + \varphi_i(1) ,$$

$$(\forall i = 1 \text{ to } N+1) . \tag{3.124}$$

The expressions of A_{ij}, and b_j corresponding to the formulas (3.97) are then obtained by identification.

▶ **Characteristic Function φ_j of a Node Strictly Interior at [0,1]**

A.7) The characteristic basis functions φ_i of the nodes of the mesh that are strictly interior at the $[0, 1]$ integration interval is now considered.

The generic equation of system (3.124) is strictly similar to the one obtained in the Dirichlet problem, (see Problem [3.1], (3.36)).

In fact, to be sure of that, it is only necessary to note that for any characteristic basis function φ_i of a node strictly interior at the $[0, 1]$ interval, the following is obtained: $\varphi_i(1) = 0$, (see Fig. 3.6).

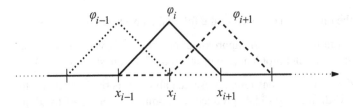

Fig. 3.6 Basis Functions φ_{i-1}, φ_i and φ_{i+1}

In that case, the nodal equ. (3.124) is then written as:

$$\sum_{j=1}^{N+1} \left[\int_0^1 \left(\varphi_j' \varphi_i' + \varphi_j \varphi_i \right) dx \right] \tilde{u}_j = \int_0^1 f \varphi_i(x) dx,$$

$$(\forall i = 1, N+1), \tag{3.125}$$

which corresponds exactly to the nodal equation (3.36) of the Dirichlet problem.

That is why the results demonstrated in the Dirichlet problem are directly reused in order to exploit them directly in the Fourier-Dirichlet problem.

▶ **Approximate Calculation of Coefficients** $A_{ij}, j = i-1, i, i+1$

a) *Approximation of coefficient* A_{ii}.

$$A_{ii} \simeq \frac{2}{h} + h. \tag{3.126}$$

b) *Approximation of coefficient* $A_{i,i-1}$.

$$A_{i,i-1} = A_{i,i+1} \simeq -\frac{1}{h}. \tag{3.127}$$

▶ **Estimation of the Second Member** b_i

$$b_i \simeq h f_i. \tag{3.128}$$

A.8) The nodal equation associated with any characteristic function φ_i of a strictly interior node is obtained by grouping results (3.126)–(3.128):

$$-\frac{\tilde{u}_{i-1} - 2\tilde{u}_i + \tilde{u}_{i+1}}{h^2} + \tilde{u}_i = f_i, \ i = 1 \text{ to } N. \tag{3.129}$$

A.9) Discretisation of the second order differential equation of the continuous problem (**CP**) by finite differences is strictly similar to the one presented in the Dirichlet problem.

This discretisation is written as:

$$-\frac{u(x_{i-1}) - 2u(x_i) + u(x_{i+1})}{h^2} + u(x_i) = f(x_i) + O(h^2), (i = 1 \text{ to } N). \tag{3.130}$$

Then, the traces $u_i \equiv u(x_i)$ of function u, having nodes x_i are replaced by the respective approximations $(\tilde{u}_i \approx u_i)$ in order to keep equality between the two members of (3.130) during the elimination of the infinitesimal $O(h^2)$.

The result of this substitution is that the finite differences scheme obtained corresponds exactly to the nodal equation (3.129) associated with any characteristic function φ_i of a node x_i strictly interior at the $[0, 1]$ interval.

Moreover, the finite differences scheme (3.129) is of the second order, given that the approximation consists in neglecting the term in $O(h^2)$ in the equation (3.130).

► **Characteristic Basis Function** φ_{N+1} **of the Node** $x_{N+1} = 1$

A.10) The nodal equation associated with the characteristic basis function φ_{N+1} of the node $x_{N+1} = 1$ is written as:

$$(\widetilde{\mathbf{VP}_{N+1}}) \quad A_{N+1,N}\tilde{u}_N + A_{N+1,N+1}\tilde{u}_{N+1} = b_{N+1} . \tag{3.131}$$

The coefficients $A_{N+1,N}$, $A_{N+1,N+1}$ and the second member b_{N+1} whose estimations are obtained below intervene in this equation (3.131).

► **Approximate Calculation of Coefficients** $A_{N+1,N}$ **and** $A_{N+1,N+1}$

a) *Approximation of coefficient* $A_{N+1,N+1}$.

$$A_{N+1,N+1} = \int_0^1 \left(\varphi_{N+1}'^2 + \varphi_{N+1}^2\right) dx + k\varphi_{N+1}^2(1) ,$$

$$= \int_{x_N}^{x_{N+1}} \left(\varphi_{N+1}'^2 + \varphi_{N+1}^2\right) dx + k ,$$

$$\simeq \left(\frac{1}{h^2} \times h\right) + \frac{h}{2}(0+1) + k ,$$

$$\simeq \frac{1}{h} + \frac{h}{2} + k . \tag{3.132}$$

It will be noticed that the characteristic property of the basis function φ_{N+1} has been used at abscissa x_{N+1}: $\varphi_{N+1}(x_{N+1}) = 1$.

b) *Approximation of coefficient* $A_{N+1,N}$.

$$A_{N+1,N} = \int_0^1 \left(\varphi_N'\varphi_{N+1}' + \varphi_N\varphi_{N+1}\right) dx + k\varphi_N(1)\varphi_{N+1}(1) ,$$

$$= \int_{Supp\ \varphi_N \cap Supp\ \varphi_{N+1}} \left(\varphi_N'\varphi_{N+1}' + \varphi_N\varphi_{N+1}\right) dx ,$$

$$\simeq -\frac{1}{h^2} \times h + \frac{h}{2}\left[(0 \times 1) + (1 \times 0)\right] \simeq -\frac{1}{h} . \tag{3.133}$$

Likewise, it will be noticed that the property of the basis function φ_N has been used: $\varphi_N(x_{N+1}) = 0$.

► **Estimation of the Second Member** b_{N+1}

Starting from:

$$b_{N+1} = \int_0^1 f\varphi_{N+1}dx + \varphi_{N+1}(1) = \int_{x_N}^{x_{N+1}} f\varphi_{N+1}dx + 1 ,$$

$$\simeq \frac{h}{2}[0 + f_{N+1}] + 1 \simeq \frac{h}{2}f_{N+1} + 1 . \tag{3.134}$$

A.11) The nodal equation associated with the basis function φ_{N+1} is written by grouping the results of (3.132)–(3.134).

This equation is written as:

$$-\frac{1}{h}\tilde{u}_N + \left[\frac{1}{h} + \frac{h}{2} + k\right]\tilde{u}_{N+1} = \frac{h}{2}f_{N+1} + 1. \tag{3.135}$$

A.12) The Fourier boundary conditions at abscissa $x_{N+1} = 1$ are now discretised using finite differences.

To achieve this, a regressive Taylor's expansion is considered at abscissa x_{N+1} expressing the solution u of the continuous problem **(CP)** at abscissa x_N according to the values of u and of its derivatives at abscissa x_{N+1}.

This expansion is written as:

$$u(x_N) = u(x_{N+1}) - hu'(x_{N+1}) + \frac{h^2}{2}u''(x_{N+1}) + O(h^3). \tag{3.136}$$

Then, supposing that the differential equation of the continuous problem **(CP)** can be written at the border of the integration domain $]0,1[$ i.e. here at abscissa x_{N+1}, the second derivative of the solution u at abscissa x_{N+1} appearing in the equ. (3.136) is replaced as below:

$$u(x_N) = u(x_{N+1}) - hu'(x_{N+1}) + \frac{h^2}{2}[u(x_{N+1}) - f(x_{N+1})] + O(h^3). \tag{3.137}$$

Moreover, the first derivative of the solution u at abscissa x_{N+1}, is expressed by using the Fourier boundary conditions:

$$u'(x_{N+1}) = 1 - ku(x_{N+1}). \tag{3.138}$$

Then equ. (3.137) takes the following form:

$$u(x_N) = u(x_{N+1}) - h[1 - ku(x_{N+1})] + \frac{h^2}{2}[u(x_{N+1}) - f(x_{N+1})] + O(h^3). \tag{3.139}$$

Then, some algebraic manipulations are operated to write equation (3.139) in the following form:

$$-\frac{1}{h}u(x_N) + \left[\frac{1}{h} + \frac{h}{2} + k\right]u(x_{N+1}) = \frac{h}{2}f_{N+1} + 1 + O(h^2). \tag{3.140}$$

The nodal equ. (3.135) associated with the basis function φ_{N+1} is then obtained, provided that the values $u(x_i)$ of the solution u to the continuous problem **(CP)** are replaced by the respective approximations $\tilde{u}_i$ in equ. (3.140):

$$-\frac{1}{h}\tilde{u}_N + \left[\frac{1}{h} + \frac{h}{2} + k\right]\tilde{u}_{N+1} = \frac{h}{2}f_{N+1} + 1. \tag{3.141}$$

3.4 Periodic Problem

3.4.1 Statement

This objective of the problem is to initiate the finite element method in a second order differential problem showing periodic boundary conditions.

Actually, interest is axed on the solutions to the following continuous problem:

Find $u \in H^2(0,1)$ which is the solution to:

$$(\mathbf{CP}) \begin{cases} -u''(x) + u(x) = f(x), 0 \leq x \leq 1, \\ u(0) = u(1), u'(0) = u'(1), \end{cases} \quad (3.142)$$

where f is a given function belonging to $L^2(0,1)$.

▶ **Variational Formulation – Theoretical Part**

1) Let v be a test function, defined from $[0,1]$ to $\mathbf{R}$, belonging to variational space V.

Show that the continuous problem (**CP**) can be expressed as a variational formulation (**VP**) under the form of:

$$a(u,v) = L(v), \quad \forall v \in V.$$

The bilinear form $a(.,.)$, the linear form $L(.)$ and the functional space V need to be determined.

2) Establish the existence and uniqueness of a weak solution of the variational problem (**VP**) in $H^1_{\text{per}}(0,1)$ defined by:

$$H^1_{\text{per}}(0,1) = \{v:]0,1[\rightarrow \mathbf{R}, \ v \in L^2(0,1), \ v' \in L^2(0,1), \ v(0) = v(1)\}.$$

3) Show that any weak solution to the variational problem (**VP**) belongs also to $H^2(0,1)$.

4) Deduce the equivalence between the strong formulation presented in $H^2(0,1)$ and the weak formulation (**VP**) considered in $H^1_{\text{per}}(0,1) \cap H^2(0,1)$.

▶ **Lagrange finite element P_1 – Numerical Part**

5) The approximation to the variational problem (**VP**) is done by Lagrange finite elements P_1. To do so, a regular mesh is introduced at interval $[0,1]$

of constant step h, such as:

$$\begin{cases} x_0 = 0, \; x_{N+1} = 1, \\ x_{i+1} = x_i + h, \; i = 0 \text{ to } N. \end{cases} \qquad (3.143)$$

The approximation space $\tilde{V}$ is now defined by:

$$\tilde{V} = \left\{ \tilde{v} \colon [0,1] \to \mathbf{R}, \; \tilde{v} \in C^0([0,1]), \; \tilde{v}|_{[x_i,x_{i+1}]} \in P_1([x_i,x_{i+1}]), \; \tilde{v}(0) = \tilde{v}(1) \right\}, \qquad (3.144)$$

where $P_1([x_i,x_{i+1}])$ denotes the polynomial space defined on $[x_i,x_{i+1}]$, of degree less than or equal to one.

– What is the dimension of $\tilde{V}$?

6) In order to numerically solve the variational problem (**VP**) by finite elements, the periodic boundary conditions bearing upon the values of the v and u functions, where $x = 0$ and $x = 1$, are temporarily "set aside".

To achieve this, the $\tilde{W}$ approximation space is introduced and defined by:

$$\tilde{W} = \left\{ \tilde{w} \colon [0,1] \longrightarrow \mathbf{R}, \tilde{w} \in C^0([0,1]), \tilde{w} \in P_1([x_i,x_{i+1}]) \right\}. \qquad (3.145)$$

– What is the dimension of $\tilde{W}$?

7) Let $\varphi_i (i = 0$ to $\dim \tilde{W} - 1)$ be the basis of $\tilde{W}$ testing $\varphi_i(x_j) = \delta_{ij}$. After expressing the approximated variational formulation of solution $\tilde{u}$ (temporarily looked for in $\tilde{W}$) associated to the variational problem (**VP**), show that when choosing:

$$\tilde{w}(x) = \varphi_i(x), (i = 0, \dim \tilde{W} - 1) \quad \text{and} \quad \tilde{u}(x) = \sum_{j=0,\,\dim \tilde{W}-1} \tilde{u}_j \varphi_j \qquad (3.146)$$

the following ($\widetilde{\textbf{VP}}$) system is obtained :

$$(\widetilde{\textbf{VP}}) \sum_{j=0,\,\dim \tilde{W}-1} A_{ij}\tilde{u}_j = b_i, \quad \forall i \in \{0,\dots, \dim \tilde{W} - 1\}, \qquad (3.147)$$

where it was stated:

$$A_{ij} = \int_0^1 (\varphi_i' \varphi_j' + \varphi_i \varphi_j)\mathrm{d}x, \quad b_i = \int_0^1 f\varphi_i \mathrm{d}x. \qquad (3.148)$$

▶ **Characteristic Function φ_i of a Node Strictly Interior at $[0,1]$**

8) Considering the mesh regularity, the generic nodal equation of the (**VP**) system associated to any basis function φ_i, $(i = 1 \dim \tilde{W} - 2)$, characteristic of a node interior at $[0,1]$, is expressed as:

$$(\widetilde{\textbf{VP}_{\textbf{Int}}}) \; A_{i,i-1}\tilde{u}_{i-1} + A_{i,i}\tilde{u}_i + A_{i,i+1}\tilde{u}_{i+1} = b_i,$$
$$(\forall i = 1 \text{ to } \dim \tilde{W} - 2). \qquad (3.149)$$

– Using the trapezium rule, calculate the 4 coefficients (A_{ij}, b_i).

9) Group the results by expressing them in a corresponding nodal equation.

10) Show that the centred finite differences scheme associated to the differential equation of the continuous problem **(CP)** is obtained again. What is its precision order?

For reminder, the trapezium quadrature formula is expressed as:

$$\int_a^b \xi(s)ds \simeq \frac{(b-a)}{2}\{\xi(a)+\xi(b)\} .$$

▶ **Characteristic Function φ_0 of the Abscissa Node $x_0 = 0$**

11) The same process is used for the basis function φ_0, characteristic of the initial node x_0. The corresponding equation of the $\widetilde{(\mathbf{VP})}$ system is then expressed as:

$$(\widetilde{\mathbf{VP_0}})\ A_{00}\tilde{u}_0 + A_{01}\tilde{u}_1 = b_0 . \tag{3.150}$$

– Using the trapezium rule, calculate the A_{00}, A_{01} and b_0 coefficients.

12) Group the results by expressing them in a corresponding nodal equation.

▶ **Characteristic Function φ_{N+1} of the Abscissa Node x_{N+1}**

13) The same process is used for the basis function φ_{N+1}, characteristic of the final node x_{N+1}. The corresponding equation to the $\widetilde{(\mathbf{VP})}$ system is then expressed as:

$$(\widetilde{\mathbf{VP_{N+1}}})\ A_{N+1,N}\tilde{u}_N + A_{N+1,N+1}\tilde{u}_{N+1} = b_{N+1} . \tag{3.151}$$

– Using the trapezium rule, calculate the $A_{N+1,N}, A_{N+1,N+1}$ and b_{N+1} coefficients.

14) Group the results by expressing them in a corresponding nodal equation.

15) Considering the periodicity properties of the nodal equations characteristic of nodes x_0 and x_{N+1}, state an algebraic equation, noted **(R)**, between the unknowns $(\tilde{u}_0, \tilde{u}_1, \tilde{u}_N)$ and the data (f_0, f_{N+1}).

16) Process a second-order discretization on the periodic boundary conditions of the continuous problem **(CP)** using the finite differences method, and show that the exact previous algebraic equation **(R)** is obtained.

3.4.2 Solution

▶ **Variational Formation – Theoretical Part**

A.1) Let v be a test function, defined from $[0,1]$ to **R**, "sufficiently regular" belonging to a functional space V. The differential equation of the continuous problem (**CP**) is multiplied by v and integrated upon $[0,1]$ interval:

$$-\int_0^1 u''v\,dx + \int_0^1 uv\,dx = \int_0^1 fv\,dx, \quad \forall v \in V. \tag{3.152}$$

The integration by parts then results in:

$$\int_0^1 u'v'dx + u'(0)v(0) - u'(1)v(1) + \int_0^1 uv\,dx = \int_0^1 fv\,dx, \quad \forall v \in V. \tag{3.153}$$

It is now demonstrated that the periodic boundary conditions bearing on the u derivative $(u'(0) = u'(1))$, can be directly injected in the integral formulation (3.153).

Thus, the result obtained is:

$$\int_0^1 (u'v' + uv)dx + u'(0)(v(0) - v(1)) = \int_0^1 fv\,dx, \quad \forall v \in V. \tag{3.154}$$

Concerning the periodic boundary conditions bearing upon u, the test function v is bound to satisfy the same boundary conditions, being:

$$v(0) = v(1). \tag{3.155}$$

Thus, the variational problem (**VP**) keeps all the information contained in the continuous problem (**CP**).

The result obtained is that u is solution to the following formal variational formulation:

$$\int_0^1 (u'v' + uv)dx = \int_0^1 fv\,dx, \quad \forall v \text{ such that: } v(0) = v(1). \tag{3.156}$$

Finally, it is shown that the Cauchy-Schwartz inequality secures, as usual, the convergence of the different integrals composing the variational formulation (3.156), if v and v' are functions belonging to $L^2(0,1)$. In other words, from then on, take the test functions v – and from there, solution u – as belonging to $H^1(0,1)$.

Considering the periodic boundary conditions (3.155) imposed in addition, the result is a variational space V defined by:

$$V \equiv H^1_{per}(0,1) = \{v :]0,1[\rightarrow \mathbf{R}, v \text{ and } v \in L^2(0,1), \ v' \in L^2(0,1), \ v(0) = v(1)\}. \tag{3.157}$$

Finally, the variational formulation (**VP**) is expressed as:

$$
(\textbf{VP}) \begin{cases} \text{Find } u \text{ belonging to } V \text{ solution to: } a(u,v) = L(v), \ \forall v \in V, \text{ where:} \\[2mm] a(u,v) \equiv \int_0^1 \left[u'(x)v'(x) + u(x)v(x) \right] dx \,, \\[2mm] L(v) \equiv \int_0^1 f(x)v(x)dx \,, \\[2mm] V \equiv H^1_{\text{per}}(0,1) \,. \end{cases}
$$

$$
(3.158)
$$

A.2) The existence and uniqueness of the variational formulation (**VP**) solution is demonstrated by applying the Lax-Milgram theorem (theorem 10), using an analogous method to the one detailed in the Dirichlet problem (see problem [3.1]).

To achieve this, the space $H^1_{\text{per}}(0,1)$ is fitted with the H^1-norm (3.17) and the aim is to prove that $H^1_{\text{per}}(0,1)$ is a closed of $H^1(0,1)$, thus conferring it with the Hilbert structure for the H^1-norm.

The sequence v_n is then considered as belonging $H^1_{\text{per}}(0,1)$ to converging for norm H^1 towards a v function of $H^1(0,1)$.

The closing property of $H^1_{\text{per}}(0,1)$ in $H^1(0,1)$ consists to establish that the limit v is also an element of $H^1(0,1)$.

It is established that (Cf. H. Brézis, [1]) if v is a function of $H^1(0,1)$, v is also a continuous function (to be exact, a continuous representative in the function class equal to v almost everywhere).

Moreover, since v_n converges towards v in $H^1(0,1)$, it is inferred that (Cf. H. Brézis, [1]) a sub-sequence v_{n_k} exists, composed of v_n in such a way that v_{n_k} simply converges towards v, for nearly any x belonging to $[0,1]$ (actually, it can be established that the convergence of v_n towards v is uniform).

Since v_n and v are "continuous" functions, it is inferred that the simple convergence occurs at any x point of the interval $[0,1]$. Thus, it is possible to express the simple convergence when $x = 0$ and when $x = 1$:

$$
\lim_{n \to +\infty} v_{n_k}(0) = v(0) \quad \text{and} \quad \lim_{n \to +\infty} v_{n_k}(1) = v(1) \,. \tag{3.159}
$$

The only step left is to evaluate the $[v(0) - v(1)]$ difference where the aim is to establish that it is equal to zero, in order to certify that limit v and sequence v_n do belong to $H^1_{\text{per}}(0,1)$:

$$
|v(0) - v(1)| = |v(0) - v_{n_k}(0) + v_{n_k}(0) - v(1)| \,, \tag{3.160}
$$

$$
\leq |v(0) - v_{n_k}(0)| + |v_{n_k}(1) - v(1)| \,. \tag{3.161}
$$

The periodicty property of the sequence v_n, $(v_n(0) = v_n(1))$, has been used and applied to the sub-series v_{n_k}.

To conclude, it suffices to perform a run at the boundary of inequality (3.61) to finally obtain:

$$v(0) = v(1), \qquad (3.162)$$

which ends the demonstration and confers a Hilbert structure to $H^1_{\text{per}}(0,1)$ together with the H^1-norm defined in (3.17), as a $H^1(0,1)$ closed vector sub-space.

The display of Lax-Milgram theorem is then performed for space $H^1_{\text{per}}(0,1)$ together with the H^1-norm, with no formal difference with the presentation of the Dirichlet problem.

Therefore, a unique solution exists to the variational formulation (**VP**) that belongs to $H^1_{\text{per}}(0,1)$.

A.3) The regularity result for the variational formulation (**VP**) solution will be obtained through the same procedures as the ones previously established for the Dirichlet problem (see Problem [3.1]).

It only needs to be fitted to the $H^1_{\text{per}}(0,1)$ functional frame.

Effectively, let u be a solution to the problem (**VP**) and the result is:

$$\int_0^1 u'(x)v'(x)\,dx = \int_0^1 [f(x) - u(x)]\,v(x)\,dx, \quad \forall v \in H^1_{\text{per}}(0,1). \qquad (3.163)$$

It is then possible to choose v in $C_0^1(]0,1[)$ which is well included in $H^1_{\text{per}}(0,1)$ and a situation of variational equations family (3.26) is obtained.

The following steps of the demonstration remain unchanged and it is inferred that the variational problem's (**VP**) solution belongs to $H^1_{\text{per}}(0,1) \cap H^2(0,1)$.

A.4) Using the same way as for the previous question, the equivalence between a weak and a strong solution is processed according to the demonstration proposed for the Dirichlet problem, while adapting it to the actual functional frame, being $H^1_{\text{per}}(0,1)$.

Effectively, let v be a variational problem's (**VP**) solution, where integration by parts leads to:

$$\int_0^1 (-u'' + u - f)v(x)\,dx + u'(1)v(1) - u'(0)v(0) = 0, \quad \forall v \in H^1_{\text{per}}(0,1), \qquad (3.164)$$

or else, since $v(0) = v(1)$,

$$\int_0^1 (-u'' + u - f)v(x)\,dx + [u'(1) - u'(0)]v(0) = 0, \quad \forall v \in H^1_{\text{per}}(0,1). \qquad (3.165)$$

Again, it is possible to choose v in $\mathcal{D}(0,1)$ (being legitimate since $\mathcal{D}(0,1) \subset H^1_{\text{per}}(0,1)$). The rest of the demonstration remains unchanged.

It is next proceeded by density and it is inferred that solution u of variational formulation (**VP**) verifies the differential equation of the continuous problem (**CP**) as a functional equation in $L^2(0,1)$.

Moreover, if the second member f belongs to $L^2(0,1) \cap C^0(]0,1[)$ then the differential equation is satisfied for any x belonging to $]0,1[$ and the solution u is the classical solution of the continuous problem (**CP**) belonging to $C^2(]0,1[)$.

▶ Lagrange Finite Element P_1 – Numerical Part

A.5) A function $\tilde{v}$ belonging to $\tilde{V}$ is a continuous function over the interval $[0,1]$ and is piecewise affine. Therefore, in the absence of periodic boundary conditions, $\tilde{V}$ would be isomorphic to $\mathbf{R}^{N+2}$.

Such an explanation will prove convincing if it is observed that a function $\tilde{v}$ of $\tilde{V}$ is completely determined provided that its values at $(N+2)$ points x_i of the mesh are fixed.

Indeed, the difference between two functions of $\tilde{V}$ inevitably corresponds to a change in one of the values of theses functions in relation to one of the nodes of the mesh $x_i, (i = 0 \text{ to } N+1)$.

The periodicity constraint of the functions $\tilde{v}$ of $\tilde{V}$ consequently leads to the loss of a degree of freedom.

In other words, the following expression is finally obtained:

$$\dim \tilde{V} = N + 1. \tag{3.166}$$

A.6) The previous question provides an immediate answer by showing that, considering the periodicity constraint, the dimension of $\tilde{W}$ is equal to $N+2$.

A.7) The approximate variational formulation $(\widetilde{\mathbf{VP}})$ is obtained by substituting the functions u and v in the variational formulation (**VP**) for the respective approximate functions $\tilde{u}$ and $\tilde{v}$.

The following is then obtained:

$$\int_0^1 \tilde{u}'\tilde{w}' \, dx + \int_0^1 \tilde{u}\tilde{w} \, dx = \int_0^1 f\tilde{w} \, dx, \quad \forall \tilde{w} \in \tilde{W}. \tag{3.167}$$

Or, again, by using particular expressions defined by (3.146):

$$\sum_{j=0,N+1} \left[\int_0^1 (\varphi_i'\varphi_j' + \varphi_i\varphi_j) \, dx \right] \tilde{u}_j = \int_0^1 f\varphi_i \, dx, \quad \forall i = 0 \text{ to } N+1. \tag{3.168}$$

This expression precisely corresponds to what needs to be expounded, provided that the quantities A_{ij} and b_i, as defined by (3.148), are introduced.

▶ Function φ_i Characteristic of a Node Strictly Interior at [0,1]

A.8) The regularity of the mesh enables the constitution of a generic analysis of the nodal equation, associated with any function φ_i, which is characteristic of the interior node x_i.

Indeed, given the support properties of basis functions φ_i, $(i = 1 \text{ to } N)$, and for a fixed value i, only the values of $j = i-1, j = i, j = i+1$ in the sum of equation (3.168) can provide non-zero contributions, (see Fig. 3.7).

This is why the approximate variational equation ($\widetilde{\text{VP}}$) is written in the form ($\widetilde{\text{VP}}_{\text{Int}}$), for all values of i varying from 1 to N.

For the remainder, the same formalism, as considered for of the Dirichlet, Neumann and Fourier-Dirichlet problems, is observed.

Hence, direct use is made of the results obtained while solving these problems for the calculation of the coefficients of matrix A_{ij}, as well as from the second member b_i.

In other words, the following formula is obtained:

$$A_{i,i} \simeq \frac{2}{h} + h, \; A_{i,i-1} = A_{i,i+1} \simeq -\frac{1}{h}, \; b_i \simeq h f_i . \tag{3.169}$$

A.9) Likewise, the nodal equation corresponding to the above mentioned coefficients is once more used directly:

$$-\frac{\tilde{u}_{i-1} - 2\tilde{u}_i + \tilde{u}_{i+1}}{h^2} + \tilde{u}_i = f_i, \quad (i = 1 \text{ to } N) . \tag{3.170}$$

A.10) Discretisation by finite differences is also applied as for the Dirichlet problem to give the following:

$$-\frac{u(x_{i-1}) - 2u(x_i) + u(x_{i+1})}{h^2} + u(x_i) = f(x_i) + O\left(h^2\right), \quad (i = 1 \text{ to } N) . \tag{3.171}$$

The traces u_i of u, at the nodes x_i, are then replaced by the approximations $\tilde{u}_i$, $(\tilde{u}_i \approx u_i)$, so as to maintain equality between the two members of (3.171) when suppressing the infinitely small $O(h^2)$.

This substitution immediately leads to the nodal equation (3.170).

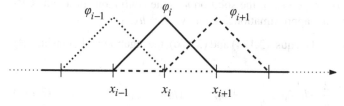

Fig. 3.7 Basis Functions φ_{i-1}, φ_i and φ_{i+1}

▶ **Basis Function φ_0 Characteristic of the Node $x_0 = 0$**

A.11) The first equation of the linear system (3.147) is now considered, that is when it corresponds to $i = 0$.

Given the support properties of basis functions φ_i, only the functions φ_0 and φ_1 can produce non-zero contributions in their integration against the function φ_0.

This is why the generic equation of system (3.147) is, in this particular case, written according to formula (3.150), namely:

$$A_{00}\tilde{u}_0 + A_{01}\tilde{u}_1 = b_0 \,. \tag{3.172}$$

Again, calculations of the coefficients A_{00}, A_{01} and b_0 have been presented in the Neumann problem (see Problem [3.2]).

The following is then obtained:

$$A_{0,0} \simeq \frac{1}{h} + \frac{h}{2}, \quad A_{0,1} \simeq -\frac{1}{h}, \quad b_0 \simeq \frac{h}{2}f_0 \,. \tag{3.173}$$

A.12) The resulting nodal equation is then written as:

$$\left[\frac{1}{h} + \frac{h}{2}\right]\tilde{u}_0 - \frac{1}{h}\tilde{u}_1 = \frac{h}{2}f_0 \,. \tag{3.174}$$

▶ **Basis Function φ_{N+1} Characteristic of the Node $x_{N+1} = 1$**

A.13) For analogous reasons to the ones previously described for the nodal equation associated with the basis function φ_0, the results obtained in the Neumann problem (see Problem [3.2]) are used directly.

Thus, the coefficients $A_{N,N+1}$, $A_{N+1,N+1}$ and b_{N+1} are given by:

$$A_{N+1,N+1} \simeq \frac{1}{h} + \frac{h}{2}, \quad A_{N+1,N} \simeq -\frac{1}{h}, \quad b_{N+1} \simeq \frac{h}{2}f_{N+1} \,. \tag{3.175}$$

A.14) The resulting nodal equation is then written as:

$$-\frac{1}{h}\tilde{u}_N + \left[\frac{1}{h} + \frac{h}{2}\right]\tilde{u}_{N+1} = \frac{h}{2}f_{N+1} \,. \tag{3.176}$$

A.15) The periodicity properties of the solution u to the continuous problem (**CP**) are now imposed into the approximation $\tilde{u}$, namely: $\tilde{u}_0 = \tilde{u}_{N+1}$.

By adding the two nodal equs. (3.174) and (3.176), the algebraic relationship (**R**) is obtained:

$$\left[\frac{2}{h} + h\right]\tilde{u}_0 - \frac{1}{h}[\tilde{u}_1 + \tilde{u}_N] = \frac{h}{2}[f_0 + f_{N+1}] \,. \tag{3.177}$$

A.16) The finite differences method, is applied to the periodic boundary conditions of the continuous problem **(CP)** by simultaneously expanding at the third order, the progressive Taylor formula at point x_0 and the regressive Taylor formula at point x_{N+1} :

$$u(x_1) = u(x_0) + hu'(x_0) + \frac{h^2}{2}u''(x_0) + O(h^3) , \tag{3.178}$$

$$u(x_N) = u(x_{N+1}) - hu'(x_{N+1}) + \frac{h^2}{2}u''(x_{N+1}) + O(h^3) . \tag{3.179}$$

It is then assumed that the solution u of the continuous problem **(CP)** is sufficiently regular so as to correctly write the differential equ. (3.142) at points x_0 and x_{N+1}:

$$u''(x_0) = u(x_0) - f(x_0) , \tag{3.180}$$
$$u''(x_{N+1}) = u(x_{N+1}) - f(x_{N+1}) . \tag{3.181}$$

The respective expressions of the second derivative (3.180) and (3.181) at points x_0 and x_{N+1} are then replaced in equs. (3.178) and (3.179):

$$u(x_1) = u(x_0) + hu'(x_0) + \frac{h^2}{2}[u(x_0) - f(x_0)] + O(h^3), \tag{3.182}$$

$$u(x_N) = u(x_{N+1}) - hu'(x_{N+1}) + \frac{h^2}{2}[u(x_{N+1}) - f(x_{N+1})] + O(h^3) . \tag{3.183}$$

The approximations $\tilde{u}_i$ are then considered and this enables one to discard the infinitely small ones while maintaining equality.

Finally, the periodicity conditions written for the sequence $\tilde{u}_i$ are applied, in order to obtain the algebraic relationship **(R)** by identifying (3.182) and (3.183):

$$\frac{h}{2}[f_0 + f_{N+1}] = \left[\frac{2}{h} + h\right]\tilde{u}_0 - \frac{1}{h}[\tilde{u}_1 + \tilde{u}_N] . \tag{3.184}$$

Chapter 4

Finite Elements in Deformable Solid Body Mechanics

4.1 Mixed Stress-Displacement Problems

4.1.1 Statement

A homogenous and isotropic elastic solid having a given Lame's Coefficient of $\lambda > 0$ and $\mu > 0$, occupies a region Ω representing an open area bordered by the three-dimensional space (x_1, x_2, x_3), (cf. Fig. 4.1).

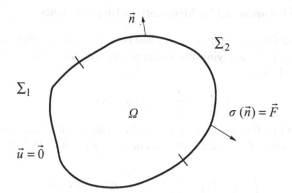

Fig. 4.1 Three-dimensional Elastic Medium

The following linear elastostatic problem is considered:

Determine $(\sigma, \mathbf{u})$ the stress and the displacement fields defined in Ω being the solution to the continuous problem (**CP**):

$$\frac{\partial \sigma_{ij}}{\partial x_j} + f_i = 0, \quad \text{in} \quad \Omega, \tag{4.1}$$

$$\mathbf{u} = \mathbf{0}, \quad \text{on} \quad \Sigma_1, \tag{4.2}$$

(**CP**) $$\sigma_{ij} n_j = F_i, \quad \text{on} \quad \Sigma_2, \tag{4.3}$$

$$\sigma_{ij} = \lambda \varepsilon_{ll}(\mathbf{u}) \delta_{ij} + 2\mu \varepsilon_{ij}(\mathbf{u}), \tag{4.4}$$

$$\varepsilon_{ij}(\mathbf{u}) = \frac{1}{2} \left[\frac{\partial u_i}{\partial x_j} + \frac{\partial u_j}{\partial x_i} \right], \tag{4.5}$$

where we have used the Einstein summation convention, also called the summation of repeated indices convention. Thus we obtain:

$$\frac{\partial \sigma_{ij}}{\partial x_j} \equiv \sum_{j=1,2,3} \frac{\partial \sigma_{ij}}{\partial x_j}, \tag{4.6}$$

$$\varepsilon_{ll}(\mathbf{u}) \equiv \text{Tr } \varepsilon(\mathbf{u}) \equiv \varepsilon_{11}(\mathbf{u}) + \varepsilon_{22}(\mathbf{u}) + \varepsilon_{33}(\mathbf{u}). \tag{4.7}$$

The mathematical notation $\text{Tr}\varepsilon(\mathbf{u})$ represents the trace of the matrix associated with the strain tensors $\varepsilon(\mathbf{u})$, of generic elements $\varepsilon_{ij}(\mathbf{u})$.

In addition, the volumetric density of efforts $\mathbf{f}$, of components f_i is given and is such that $f_i, (i = 1, 2, 3)$, belonging to $L^2(\Omega)$.

Similarly, the surface density of efforts $\mathbf{F}$, of F_i components is given and belongs to $L^2(\Sigma_2)$.

▶ Variational Formulation for Mixed Stress-Displacements

1) Let $\mathbf{v}$ be a test field (or virtual) vectors that is "sufficiently regular" and consisting of v_i components. By applying the symmetry of the stress tensor σ to solve the problem (**CP**), show that:

$$\int_{\Omega} \sigma_{ij} \frac{\partial v_i}{\partial x_j} = \int_{\Omega} \sigma_{ij} \varepsilon_{ij}(\mathbf{v}). \tag{4.8}$$

2) Infer from it that for any displacement fields $\mathbf{v}$, null on the border of Σ_1, a variational formulation (**VP**) associated to the continuous problem (**CP**) may be written as:

$$\int_{\Omega} \sigma_{ij} \varepsilon_{ij}(\mathbf{v}) = \int_{\Omega} f_i v_i + \int_{\Sigma_2} F_i v_i, \quad \forall \mathbf{v} \in V. \tag{4.9}$$

3) After having eliminated the stress tensor from the variational formulation (4.9), write the displacement variational formulation (VP)$_u$ as following:

Find $\mathbf{u} \in V$ being the solution of: $a(\mathbf{u}, \mathbf{v}) = L(\mathbf{v}), \quad \forall \mathbf{v} \in V$. $\tag{4.10}$

– Specify the functional V space.

4) Justify the existence of a minimisation problem (**MP**) that is equivalent to the variational formulation (**VP**)$_u$.

▶ Variational Formulation Associated with Navier's Equation

All along the problem, the material's geometry is restricted to a two-dimensional (x_1, x_2) situation.

It is proposed to study the mechanical system whose geometry is defined by the square $\Omega \equiv]0,1[\times]0,1[$.

In addition, the border Σ_2 of $\partial\Omega$ is made up of the segment that is $x_2 = 0$ and the border Σ_1 corresponds to the supplementary part of Σ_2 in $\partial\Omega$. In other words, Σ_1 consists of three other sides of the square Ω (cf. Fig. 4.2).

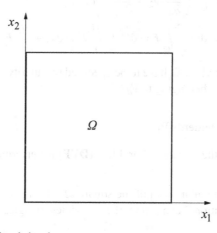

Fig. 4.2 Elastic Square $]0,1[\times]0,1[$

5) Then show that the continuous problem **(CP)** may be solved as displacements using Navier's Equation written in the vectorial form:

$$(\lambda + \mu)\overrightarrow{\text{grad}}(\text{div } \mathbf{u}) + \mu\Delta\mathbf{u} + \mathbf{f} = 0, \quad \text{in } \Omega. \tag{4.11}$$

6) Infer that the continuous problem **(DCP)**, of unknown displacements $\mathbf{u}$ has the structure shown below:

$$\textbf{(DCP)} \quad \begin{cases} (\lambda + \mu)\overrightarrow{\text{grad}}(\text{div } \mathbf{u}) + \mu\Delta\mathbf{u} + \mathbf{f} = 0, \text{ in } \Omega, & (4.12) \\[2mm] \mathbf{u} = 0, \text{ on } \Sigma_1, & (4.13) \\[2mm] \dfrac{\partial u_1}{\partial x_2}(x_1,0) + \dfrac{\partial u_2}{\partial x_1}(x_1,0) = -\dfrac{F_1}{\mu}, & (4.14) \\[2mm] \lambda\dfrac{\partial u_1}{\partial x_1}(x_1,0) + (\lambda + 2\mu)\dfrac{\partial u_2}{\partial x_2}(x_1,0) = -F_2, & (4.15) \end{cases}$$

where u_1 and u_2 are the components of the displacement field $\mathbf{u}$.

7) Let $\mathbf{v}$ be an arbitrary field of vectors belonging to V_d. Show that a variational formulation **(DVP)** associated to the continuous problem **(DCP)** is written as:

(DVP)

$$
\begin{cases}
\text{Find } \mathbf{u} \in V_d \text{ the solution of:} \quad a(\mathbf{u},\mathbf{v}) = L(\mathbf{v}), \quad \forall \mathbf{v} \in V_d, & (4.16) \\[2mm]
\text{with:} \\[2mm]
a(\mathbf{u},\mathbf{v}) = \mu \int_\Omega [\boldsymbol{\nabla} u_1 \boldsymbol{\nabla} v_1 + \boldsymbol{\nabla} u_2 \boldsymbol{\nabla} v_2]\, \mathrm{d}\Omega \\[2mm]
\qquad\qquad + (\lambda + \mu) \int_\Omega \operatorname{div} \mathbf{u} \cdot \operatorname{div} \mathbf{v}\, \mathrm{d}\Omega & (4.17) \\[2mm]
\qquad\qquad + \mu \int_0^L \left(\dfrac{\partial u_1}{\partial x_1} v_2 - \dfrac{\partial u_2}{\partial x_1} v_1 \right) \mathrm{d}x_1, \\[2mm]
L(\mathbf{v}) = \displaystyle\int_\Omega f_1 v_1\, \mathrm{d}\Omega + \int_\Omega f_2 v_2\, \mathrm{d}\Omega + \int_0^L F_1 v_1\, \mathrm{d}x_1 + \int_0^L F_2 v_2\, \mathrm{d}x_1. & (4.18)
\end{cases}
$$

The nature of the space V_d will have to be specified (regularity and boundary conditions of the functions v belonging to V_d).

▶ **Lagrange Finite Elements P_1**

8) Approximation of the variational problem **(DVP)** is performed using Lagrange's finite elements $\mathbf{P}_1$.

To achieve this, a regular mesh of the square Ω (cf. Fig. 4.3) is introduced with the help of a triangulation $\mathfrak{I}$ consisting of T_k isosceles triangles having sides h, such that:

$$
\begin{cases}
x_1^{(0)} = x_2^{(0)} = 0,\; x_1^{(N+1)} = x_2^{(N+1)} = L, \\[2mm]
x_1^{(i+1)} = x_2^{(i)} + h,\; x_2^{(i+1)} = x_2^{(i)} + h,\; (i = 0 \text{ to } N).
\end{cases} \tag{4.19}
$$

Now the approximation space $\tilde{V}$ is defined by:

$$
\tilde{V} = \left\{ \tilde{\mathbf{v}}\colon \Omega \to \mathbf{R}^2,\, \tilde{\mathbf{v}} \in \left[C^0(\Omega) \right]^2,\, \tilde{v}_i|_{T_k} \in P_1(T_k),\, \tilde{\mathbf{v}} = \mathbf{0} \text{ on } \Sigma_1 \right\}, \tag{4.20}
$$

where $P_1(T_k)$ refers to the polynomial space specified on the generic triangle T_k, having a degree less than or equal to 1 relative to the two variables (x_1, x_2) of the space.

– What is the dimension of $\tilde{V}$?

▶ **Approximate Variational Formulation**

9) Assume $P = N(N+1)$. Thus φ_i, $(i = 1 \text{ to } P)$ being the classical basis functions used for P_1 finite elements that satisfies:

$$
\varphi_i(M_j) = \delta_{ij}.
$$

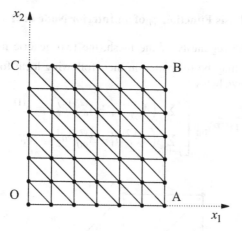

Fig. 4.3 Mesh for Finite Element P_1

Having written the approximate variational formulation $(\widetilde{\mathbf{DVP}})$ of the solution $\tilde{\mathbf{u}}$, associated to the variational problem **(DVP)**, show that by writing $\tilde{\mathbf{u}}$ in the form of:

$$\tilde{\mathbf{u}} = \left(\sum_{j=1,P} \tilde{u}_1{}^j \varphi_j, \sum_{j=1,P} \tilde{u}_2{}^j \varphi_j \right), \qquad (4.21)$$

and by successively choosing $\tilde{\mathbf{v}}$ in the form of:

$$\tilde{\mathbf{v}} = (\varphi_i, 0), \quad \text{then} \quad \tilde{\mathbf{v}} = (0, \varphi_i), \quad (i = 1, P), \qquad (4.22)$$

The following linear system is obtained:

$$\widetilde{\mathbf{(DVP)}} \begin{bmatrix} A_{ij}^{(1,1)} \tilde{u}_{1j} + A_{ij}^{(2,1)} \tilde{u}_{2j} = b_i^{(1)}, \\ A_{ij}^{(1,2)} \tilde{u}_{1j} + A_{ij}^{(2,2)} \tilde{u}_{2j} = b_i^{(2)}, \quad (\forall i = 1, P), \end{bmatrix} \qquad (4.23)$$

with:

$$A_{ij}^{(1,1)} = \mu \int_\Omega \nabla \varphi_i \cdot \nabla \varphi_j + (\lambda + \mu) \int_\Omega \frac{\partial \varphi_i}{\partial x_1} \frac{\partial \varphi_j}{\partial x_1}, \qquad (4.24)$$

$$A_{ij}^{(2,1)} = (\lambda + \mu) \int_\Omega \frac{\partial \varphi_i}{\partial x_1} \frac{\partial \varphi_j}{\partial x_2} - \mu \int_0^L \varphi_i \frac{\partial \varphi_j}{\partial x_1} \, dx_1, \qquad (4.25)$$

$$A_{ij}^{(1,2)} = (\lambda + \mu) \int_\Omega \frac{\partial \varphi_i}{\partial x_2} \frac{\partial \varphi_j}{\partial x_1} + \mu \int_0^L \varphi_i \frac{\partial \varphi_j}{\partial x_1} \, dx_1, \qquad (4.26)$$

$$A_{ij}^{(2,2)} = \mu \int_\Omega \nabla \varphi_i \cdot \nabla \varphi_j + (\lambda + \mu) \int_\Omega \frac{\partial \varphi_i}{\partial x_2} \frac{\partial \varphi_j}{\partial x_2}, \qquad (4.27)$$

$$b_i^{(1)} = \int_\Omega f_1 \varphi_i + \int_0^L F_1 \varphi_i \, dx_1,$$

$$b_i^{(2)} = \int_\Omega f_2 \varphi_i + \int_0^L F_2 \varphi_i \, dx_1. \qquad (4.28)$$

▶ **Characteristic Basis Function** φ_i **of an Interior Node at** Ω

10) Considering the regularity of the mesh, the two generic nodal equations of the $\widetilde{(\textbf{DVP})}$ system may be re-written in their following local forms using the local numeration shown below:

$$
\widetilde{(\textbf{DVP})}_{\textbf{Int}} \left[\begin{array}{l} \displaystyle\sum_{j=0,6} \left(A_{o,j}^{(1,1)} \tilde{u}_1^j + A_{o,j}^{(2,1)} \tilde{u}_2^j \right) = b_o^{(1)} , \\[4mm] \displaystyle\sum_{j=0,6} \left(A_{o,j}^{(1,2)} \tilde{u}_1^j + A_{o,j}^{(2,2)} \tilde{u}_2^j \right) = b_o^{(2)} . \end{array} \right. \tag{4.29}
$$

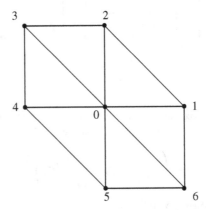

Fig. 4.4 Local Numeration Associated to a Node Strictly Interior to Ω

In the rest of the problem, only the first nodal equation of the system will explicitly be dealt with (4.29), i. e., corresponding to the particular choice of $\tilde{\textbf{v}} = (\varphi_i, 0)$, knowing that the second nodal equation is handled in an analogous manner.

– Calculate the 14 coefficients $A_{0,j}^{(1,1)}, A_{0,j}^{(2,1)}, (j = 0$ to $6)$, as well as the second member $b_0^{(1)}$.

11) Bring the results together by writing the corresponding nodal equation of the $\widetilde{(\textbf{DVP})}_{\textbf{Int}}$ system.

12) Show that the pattern with centered finite-differences is obtained again and is associated to the first partial differential equation of the continuous problem **(DCP)**. What is the order of its precision?

It is pointed out that the trapezium quadrature formula applied to a triangle T_{123}, whose vertices are given as A_1, A_2 and A_3 is written as:

$$
\iint_{T_{123}} f(x,y) \, dx \, dy \simeq \frac{\text{Area } (T_{123})}{3} \sum_{i=1,2,3} f(A_i) , \tag{4.30}
$$

where the area of triangle T_{123} is denoted by Area(T_{123}).

▶ **Characteristic Basis Function φ_i of an Interior Node at Σ_2**

13) By keeping the local numeration (cf. Fig. 4.5) and by considering the geometric specificities of the border Σ_2, the nodal equations corresponding to the characteristic basis functions of the border nodes Σ_2 are written as:

$$
\widetilde{(\mathbf{DVP})}_{\Sigma_2} \quad
\left[
\begin{array}{l}
\displaystyle\sum_{j=0,4} \left(A_{0,j}^{(1,1)} \tilde{u}_1^j + A_{0,j}^{(2,1)} \tilde{u}_2^j \right) = b_0^{(1)}\,, \\[4mm]
\displaystyle\sum_{j=0,4} \left(A_{0,j}^{(1,2)} \tilde{u}_1^j + A_{0,j}^{(2,2)} \tilde{u}_2^j \right) = b_0^{(2)}\,.
\end{array}
\right.
\tag{4.31}
$$

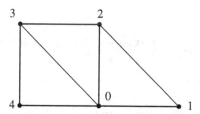

Fig. 4.5 Local Numeration Associated to a Node Found Inside Σ_2

– Calculate the 10 coefficients $A_{0,j}^{(1,1)}, A_{0,j}^{(2,1)}, (j = 0 \text{ in } 4)$, as well as the second member $b_0^{(1)}$.

14) Bring the results together by writing the corresponding nodal equation of the $\widetilde{(\mathbf{DVP})}_{\Sigma_2}$ system.

15) Show that the pattern with centered finite-differences of the second order, associated to the system of the two equations defining the stress on Σ_2 (4.14)–(4.15) of the (**DCP**) continuous problem, is obtained again.

4.1.2 Solution

▶ **Variational Formulation for Mixed Stress-Displacements**

A.1) Let **v** be a field of test vectors having all the necessary regularity for carrying out integration calculations that will be developed in the questions that follow.

Formula (4.8) is obtained by inserting the strain tensors $\varepsilon(\mathbf{v})$ using the left integral of formula (4.8).

To achieve this, the integral of the stress tensors σ against the complementary partial derivative of $\dfrac{\partial v_j}{\partial x_i}$ appearing in the definition of the strain tensors $\varepsilon(\mathbf{v})$ (4.5) is derived:

$$\int_\Omega \sigma_{ij} \frac{\partial v_j}{\partial x_i} = \int_\Omega \sigma_{ji} \frac{\partial v_i}{\partial x_j} = \int_\Omega \sigma_{ij} \frac{\partial v_i}{\partial x_j}, \quad \text{(symmetry of the tensor } \sigma\text{)}. \qquad (4.32)$$

This brings the deduction:

$$\int_\Omega \sigma_{ij} \left[\frac{\partial v_j}{\partial x_i} + \frac{\partial v_i}{\partial x_j} \right] = 2 \int_\Omega \sigma_{ij} \frac{\partial v_i}{\partial x_j}. \qquad (4.33)$$

Formula (4.8) is obtained if care is taken to introduce the definition (4.5) of the strain tensors $\varepsilon(\mathbf{v})$.

A.2) In order to produce a variational formulation (**VP**), the partial differential equation (4.1) is multiplied by the v_i component of the test displacement field **v** and is integrated on the Ω domain:

$$\int_\Omega \frac{\partial \sigma_{ij}}{\partial x_j} v_i + \int_\Omega f_i v_i = 0. \qquad (4.34)$$

In addition, Green's formula is used to transform the first member of (4.34).

This operation yields the equation below:

$$-\int_\Omega \sigma_{ij} \frac{\partial v_i}{\partial x_j} + \int_{\partial\Omega} \sigma_{ij} n_j v_i + \int_\Omega f_i v_i = 0. \qquad (4.35)$$

Then, formula (4.8) and the stress boundary conditions (4.3) are considered and these yield:

$$\int_\Omega \sigma_{ij} \varepsilon_{ij}(\mathbf{v}) = \int_{\Sigma_1} \sigma_{ij} n_j v_i + \int_\Omega f_i v_i + \int_{\Sigma_2} F_i v_i. \qquad (4.36)$$

Since it is impossible to integrate boundary conditions (4.2) bearing on the displacement field solution **u** into formula (4.36), the variational space V is constructed in such a way that its elements **v** are identically nil on Σ_1 so that the solution being a particular element of V be also nil on the Σ_1 border of Ω.

The formula then becomes:

$$\int_{\Omega} \sigma_{ij}\varepsilon_{ij}(\mathbf{v}) = \int_{\Omega} f_i v_i + \int_{\Sigma_2} F_i v_i, \quad \forall \mathbf{v} \in V = \{\mathbf{v}/\mathbf{v} = \mathbf{0} \text{ on } \Sigma_1\} . \qquad (4.37)$$

A.3) At this stage the behaviour law of the material (4.35) is used so as to replace the stress tensor σ by the strain tensor $\varepsilon(\mathbf{u})$ and subsequently by the displacement field solution $\mathbf{u}$.

Equation (4.37) then leads to the variational formulation $(\mathbf{VP})_u$:

$$(\mathbf{VP})_u \quad \left[\begin{array}{l} \text{Find } \mathbf{u} \in V, \text{ the solution to: } \forall \mathbf{v} \in V = \left\{ \mathbf{v}/\mathbf{v} = \mathbf{0} \text{ on } \Sigma_1 \right\} : \\ \int_{\Omega} [\lambda \varepsilon_{ll}(\mathbf{u})\delta_{ij} + 2\mu\varepsilon_{ij}(\mathbf{u})] \varepsilon_{ij}(\mathbf{v}) = \int_{\Omega} f_i v_i + \int_{\Sigma_2} F_i v_i , \end{array} \right. \qquad (4.38)$$

or else, by using the properties of Krönecker's symbol:

$$(\mathbf{VP})_u \quad \left[\begin{array}{l} \qquad\qquad a(\mathbf{u}, \mathbf{v}) = L(\mathbf{v}) , \quad \forall \mathbf{v} \in V , \qquad\qquad (4.39) \\ \text{with:} \\ a(\mathbf{u}, \mathbf{v}) \equiv \int_{\Omega} [\lambda \varepsilon_{ll}(\mathbf{u})\varepsilon_{mm}(\mathbf{v}) + 2\mu\varepsilon_{ij}(\mathbf{u})\varepsilon_{ij}(\mathbf{v})] , \qquad (4.40) \\ L(\mathbf{v}) \equiv \int_{\Omega} f_i v_i + \int_{\Sigma_2} F_i v_i . \qquad\qquad\qquad\qquad (4.41) \end{array} \right.$$

▶ **Properties of the space functions V**

As should be, so as to guarantee the convergence of the different integrals contributing in formulation (4.40)–(4.41), the necessary conditions bearing on the regularity of the field of vectors $\mathbf{v}$ belonging to V will now be introduced. On several occasions (cf. Dirichlet's Problem [3.1], Neumann's [3.2], etc.), it has been shown how the Cauchy-Schwartz inequality enabled the control of integrals having the form of those contributing in the expression of $L(\mathbf{v})$, given that the densities f and F belong to L^2. This is precisely what has been assumed at the beginning of this problem.

Concerning the two terms that are under the integral of the quantity $a(\mathbf{u}, \mathbf{v})$, it is seen that only linear combinations of partial derivatives of the first order appear in the expression of $a(\mathbf{u}, \mathbf{v})$.

The Cauchy-Schwartz inequality thus enables one to write in a generic manner:

$$\left| \int_{\Omega} \frac{\partial u_i}{\partial x_j} \frac{\partial v_k}{\partial x_l} \right| \leq \left[\int_{\Omega} \left| \frac{\partial u_i}{\partial x_j} \right|^2 \right]^{1/2} \left[\int_{\Omega} \left| \frac{\partial v_k}{\partial x_l} \right|^2 \right]^{1/2} . \qquad (4.42)$$

It is observed that it suffices to consider the field of vectors in a functional space such as the first partial derivatives belong to $L^2(\Omega)$.

Thus, given that it was observed earlier that the v_i components of the field $\mathbf{v}$ should also belong to $L^2(\Omega)$, the definition below is finally drawn:

$$V = \left\{ \mathbf{v}: \Omega \to \mathbf{R}^2 / v_i \in L^2(\Omega) \, , \, \frac{\partial v_i}{\partial x_j} \in L^2(\Omega) \text{ and } v_i = 0 \text{ on } \Sigma_1 \right\} . \qquad (4.43)$$

The above represents the Sobolev space $H^1(\Omega) \times H^1(\Omega)$ for the vector fields in $\mathbf{R}^2$ that, in addition, are nil on Σ_1.

A.4) The existence of an equivalent minimisation problem **(MP)** associated to the variational problem **(VP)**$_u$ is the result of the following properties of forms $a(.,.)$ and $L(.)$, as respectively defined by (4.40) and (4.41).

1. $a(.,.)$ and $L(.)$ are bilinear and linear forms respectively.

2. $a(.,.)$ has a symmetrical form.

3. $a(.,.)$ has a positive form.

In general, more of the $a(.,.)$ form is required in case the Lax-Milgram theorem (cf. Theorem 10) has to be applied.

In this case, it is necessary to establish that form $a(.,.)$ is V-elliptic. From then on, a V-elliptical form is automatically positive.

In addition, D. Euvrard [4] demonstrates the equivalence between **(MP)** and **(VP)**$_u$ problems under previously inventoried conditions.

▶ **Variational Formulation Associated with Navier's Equation**

A.5) Navier's equation is obtained by replacing the stress tensor σ in equation (4.1) by its expression supplied by Hook's Law (4.4) as a function of the strain tensors and consequently as a function of the displacement field solution $\mathbf{u}$.

Thus the formula below is obtained:

$$\frac{\partial}{\partial x_j} [\lambda \, \varepsilon_{ll}(\mathbf{u}) \delta_{ij} + 2\mu \varepsilon_{ij}(\mathbf{u})] + f_i = 0 \, , \quad \forall i = 1 \text{ to } 3 \, , \qquad (4.44)$$

or:

$$\lambda \frac{\partial \varepsilon_{ll}(\mathbf{u})}{\partial x_j} \delta_{ij} + 2\mu \frac{\partial \varepsilon_{ij}(\mathbf{u})}{\partial x_j} + f_i = 0 \, , \quad \forall i = 1 \text{ to } 3 \, . \qquad (4.45)$$

The properties of Krönecker's symbol on one side and permutation of the second partial derivative in the case of a "sufficiently regular" solution on the other side, enables the re-writing of (4.45) by using the definition of the linear strain tensor (4.5) in the form below:

$$\lambda \frac{\partial}{\partial x_i} \text{div}(\mathbf{u}) + \mu \frac{\partial^2 u_i}{\partial x_j \partial x_j} + \mu \frac{\partial}{\partial x_i} \left(\frac{\partial u_j}{\partial x_j} \right) + f_i = 0, \quad \forall i = 1 \text{ to } 3. \tag{4.46}$$

And finally:

$$(\lambda + \mu) \frac{\partial}{\partial x_i} (\text{div} \, \mathbf{u}) + \mu \Delta u_i + f_i = 0, \quad \forall i = 1 \text{ to } 3. \tag{4.47}$$

A.6) Formulation of the displacements continuous problem (**DCP**) is immediate. To achieve this, it is only necessary to replace the equilibrium equation (4.1) of the continuous problem (**CP**) by Navier's equation (4.47), on one side, and to express the conditions to the stress (4.3) on Σ_2 as a function of the displacement field solution $\mathbf{u}$.

Thus, given that the Σ_2 border corresponds to the $x_2 = 0$ segment, the normal exterior vector is $-\mathbf{x}_2$.

Subsequently, the boundary conditions (4.3) are written as following:

$$\begin{pmatrix} \sigma_{11}(x_1, 0) & \sigma_{12}(x_1, 0) \\ \sigma_{21}(x_1, 0) & \sigma_{22}(x_1, 0) \end{pmatrix} \begin{pmatrix} 0 \\ -1 \end{pmatrix} = \begin{pmatrix} F_1(x_1) \\ F_2(x_1) \end{pmatrix}, \tag{4.48}$$

so,

$$\begin{cases} -\sigma_{12}(x_1, 0) = F_1(x_1), \\ -\sigma_{22}(x_1, 0) = F_2(x_1). \end{cases} \tag{4.49}$$

Thus, by expressing the stress components σ_{12} and σ_{22} as a function of the components of the displacement field solution $\mathbf{u}$, equations (4.14)–(4.15) are obtained from the formulation of the continuous problem (**DCP**).

A.7) Now a variational formulation (**DVP**) corresponding to the continuous problem (**DCP**) is established.

To achieve this, consider an arbitrary field of vectors $\mathbf{v}$ belonging to a functional space V_d that will be specified later.

Then the Navier's equation is multiplied in a scalar manner by the v_i component yielding:

$$(\lambda + \mu) \int_\Omega \frac{\partial}{\partial x_i} \text{div} \, \mathbf{u} \cdot v_i + \mu \int_\Omega \Delta u_i \cdot v_i + \int_\Omega f_i \cdot v_i = 0. \tag{4.50}$$

Green's formula is then used to obtain:

$$-(\lambda + \mu) \int_\Omega \text{div} \, \mathbf{u} \cdot \frac{\partial v_i}{\partial x_i} + (\lambda + \mu) \int_{\partial \Omega} \text{div} \, \mathbf{u} \cdot v_i n_i + \dots$$
$$-\mu \int_\Omega \nabla u_i \cdot \nabla v_i + \mu \int_{\partial \Omega} \frac{\partial u_i}{\partial n} \cdot v_i + \int_\Omega f_i \cdot v_i = 0. \tag{4.51}$$

It is then assumed that the displacement field $\mathbf{v}$ homogeneously satisfies the Dirichlet condition on $\Sigma_1 : \mathbf{v} = \mathbf{0}$ on Σ_1.

The two border integrals bearing on $\partial\Omega$ and contributing to the variational equation (4.51) are evaluated in the following manner:

$$\int_{\partial\Omega} \operatorname{div} \mathbf{u} \cdot v_i n_i = \int_0^L \operatorname{div} \mathbf{u}(-v_2) = -\int_0^L \left[\frac{\partial u_1}{\partial x_1} v_2 + \frac{\partial u_2}{\partial x_2} v_2 \right], \tag{4.52}$$

$$\int_{\partial\Omega} \frac{\partial u_i}{\partial n} \cdot v_i = \int_0^L -\frac{\partial u_i}{\partial x_2} \cdot v_i = -\int_0^L \left[\frac{\partial u_1}{\partial x_2} \cdot v_1 + \frac{\partial u_2}{\partial x_2} \cdot v_2 \right]. \tag{4.53}$$

Then the results of equation (4.52)–(4.53) are injected in formula (4.51) to obtain:

$$(\lambda + \mu) \int_\Omega \operatorname{div} \mathbf{u} \operatorname{div} \mathbf{v} + (\lambda + \mu) \int_0^L (\operatorname{div} \mathbf{u}) v_2 + \dots$$

$$\dots + \mu \int_\Omega \nabla u_i \cdot \nabla v_i + \mu \int_0^L \left[\frac{\partial u_1}{\partial x_2} \cdot v_1 + \frac{\partial u_2}{\partial x_2} \cdot v_2 \right] = \int_\Omega f_i \cdot v_i. \tag{4.54}$$

Boundary conditions on Σ_2, written as a function of components (u_1, u_2) of the vector field solution $\mathbf{u}$ as well as of the density of efforts $\mathbf{F}$ given on the border of Σ_2, enable the re-writing of the variational formula (4.54) following some elementary reorganization in the form of:

(DVP)
$$\left[\begin{array}{l} \text{Find } \mathbf{u} = (u_1, u_2) \in V \text{ such that: } a(\mathbf{u}, \mathbf{v}) = L(\mathbf{v}), \ \forall v \in V \text{ where:} \\[2mm] a(\mathbf{u}, \mathbf{v}) = \mu \int_\Omega \nabla u_1 \cdot \nabla v_1 + \mu \int_\Omega \nabla u_2 \cdot \nabla v_2 + (\lambda + \mu) \int_\Omega \operatorname{div} \mathbf{u} \cdot \operatorname{div} \mathbf{v} \\[4mm] \qquad + \mu \int_0^L \left(\frac{\partial u_1}{\partial x_1} v_2 - \frac{\partial u_2}{\partial x_1} v_1 \right), \hfill (4.55) \\[4mm] L(\mathbf{v}) = \int_\Omega f_1 \, v_1 + \int_\Omega f_2 \cdot v_2 + \int_0^L F_1 \cdot v_1 + \int_0^L F_2 \cdot v_2. \hfill (4.56) \end{array} \right.$$

▶ Lagrange Finite Elements P_1

A.8) The approximation space is nothing else than the Cartesian product of the classical approximation space by finite element P_1, for a scalar approximation within a square, by itself.

In other words, the dimension of $\tilde{V}$ is equal to:

$$\dim \tilde{V} = 2N(N+1) = [N(N+1) + N(N+1)],$$

because each component of the vector field solution $\mathbf{u}$ is approximated by a scalar function of approximation $\tilde{u}_i$, which describes the classical approximation space of the finite elements P_1, mentioned above.

▶ Approximate Variational Formulation

A.9) The approximate variational formulation $(\widetilde{\textbf{DVP}})$ is immediately obtained after replacing $(\textbf{u}, \textbf{v})$ in the variational formulation $(\textbf{DVP})$ by the corresponding approximation fields $(\tilde{\textbf{u}}, \tilde{\textbf{v}})$.

Moreover, in order to get the two corresponding generic equations for each component of the approximated field displacement $\tilde{\textbf{u}}$, there has to be the test vector field $\tilde{\textbf{v}}$ on one part, expressed as $\tilde{\textbf{v}} = (\varphi_i, 0)$, followed by $\tilde{\textbf{v}} = (0, \varphi_i)$.

Therefore, when $\tilde{\textbf{v}} = (\varphi_i, 0)$, we obtain:

$$a[(u_1, u_2), (\varphi_i, 0)] = \sum_j \left(\mu \int_\Omega \nabla \varphi_i \cdot \nabla \varphi_j + (\lambda + \mu) \int_\Omega \frac{\partial \varphi_i}{\partial x_1} \frac{\partial \varphi_j}{\partial x_1} \right) \tilde{u}_1^j \qquad (4.57)$$

$$+ \sum_j \left((\lambda + \mu) \int_\Omega \frac{\partial \varphi_i}{\partial x_1} \frac{\partial \varphi_j}{\partial x_2} - \mu \int_0^L \varphi_i \frac{\partial \varphi_j}{\partial x_1} \right) \tilde{u}_2^j, \qquad (4.58)$$

$$L[(\varphi_i, 0)] = \int_\Omega f_1 \varphi_i + \int_0^L F_1 \varphi_i. \qquad (4.59)$$

In the same manner, by choosing $\tilde{\textbf{v}} = (0, \varphi_i)$, we obtain:

$$a[(u_1, u_2), (0, \varphi_i)] = \sum_j \left((\lambda + \mu) \int_\Omega \frac{\partial \varphi_i}{\partial x_2} \frac{\partial \varphi_j}{\partial x_1} + \mu \int_0^L \varphi_i \frac{\partial \varphi_j}{\partial x_1} \right) \tilde{u}_1^j$$

$$+ \sum_j \left(\mu \int_\Omega \nabla \varphi_i \cdot \nabla \varphi_j + (\lambda + \mu) \int_\Omega \frac{\partial \varphi_i}{\partial x_2} \frac{\partial \varphi_j}{\partial x_2} \right) \tilde{u}_2^j, \qquad (4.60)$$

$$L[(0, \varphi_i)] = \int_\Omega f_2 \varphi_i + \int_0^L F_2 \varphi_i. \qquad (4.61)$$

Thus, the equations system (4.29) is obtained by simple identification.

▶ Basis function φ_i Characteristic of an Node Interior at Ω

A.10) The calculations that will be presented are based on the approximation of finite elements $\textbf{P}_1$ of the Laplace-Dirichlet problem in a square discussed in D. Euvrard's book [4].

To be more precise, it is the evaluations of the integrals by triangle that will be directly used in the solution below.

This was done by systematic exploitation of the local numeration presented in the statement, (cf. Fig. 4.4).

▶ Calculation of the Coefficients $A_{oj}^{(1,1)}$

$$A_{oj}^{(1,1)} = \mu \int_\Omega \nabla \varphi_o \cdot \nabla \varphi_j + (\lambda + \mu) \int_\Omega \frac{\partial \varphi_o}{\partial x_1} \frac{\partial \varphi_j}{\partial x_1}. \qquad (4.62)$$

a) Calculation of the coefficient $A_{oo}^{(1,1)}$

$$A_{oo}^{(1,1)} = \mu \int_{\text{Support } \varphi_o} |\nabla \varphi_o|^2 + (\lambda + \mu) \int_{\Omega} \left(\frac{\partial \varphi_o}{\partial x_1} \right)^2 . \tag{4.63}$$

However,

$$\int_{\text{Support } \varphi_o} |\nabla \varphi_o|^2 = \int_{012} |\nabla \varphi_o|^2 + \int_{023} |\nabla \varphi_o|^2$$

$$+ \int_{034} |\nabla \varphi_o|^2 + \int_{045} |\nabla \varphi_o|^2 + \int_{056} |\nabla \varphi_o|^2 + \int_{061} |\nabla \varphi_o|^2 ,$$

$$= \frac{h^2}{2} \left\{ \frac{2}{h^2} + \frac{1}{h^2} + \frac{1}{h^2} + \frac{2}{h^2} + \frac{1}{h^2} + \frac{1}{h^2} \right\} = 4 . \tag{4.64}$$

Moreover,

$$\int_{\Omega} \left(\frac{\partial \varphi_o}{\partial x_1} \right)^2 = 2 \times \left(\frac{1}{h^2} \right) \times \left(\frac{h^2}{2} \right) + 0 + \left(\frac{1}{h^2} \right) \times \left(\frac{h^2}{2} \right) \times 2 + 0 . \tag{4.65}$$

By grouping the results of (4.64) and (4.65), the expression of the coefficient $A_{oo}^{(1,1)}$ is obtained with the help of the definition (4.62):

$$A_{oo}^{(1,1)} = 4\mu + 2(\lambda + \mu) . \tag{4.66}$$

By proceeding in an analogous manner for the other coefficients $A_{oj}^{(1,1)}$, given that the calculations are modified to suit the intersection of the basis functions supports φ_j, the formula below is obtained:

b) Calculation of the coefficient $A_{oj}^{(1,1)}, (j = 1 \text{ to } 3)$

$$A_{o1}^{(1,1)} = -\mu - (\lambda + \mu) , \tag{4.67}$$

$$A_{o2}^{(1,1)} = -\mu , \tag{4.68}$$

$$A_{o3}^{(1,1)} = 0 . \tag{4.69}$$

The other coefficients $A_{oj}^{(1,1)}, (j = 4 \text{ to } 6)$, are obtained by the properties of geometrical symmetry and of invariant translation of the mesh in the two directions that generate the plan.

Then the following formula is thus obtained:

$$A_{o4}^{(1,1)} = A_{o1}^{(1,1)}, \tag{4.70}$$

$$A_{o5}^{(1,1)} = A_{o2}^{(1,1)}, \tag{4.71}$$

$$A_{o6}^{(1,1)} = A_{o3}^{(1,1)}. \tag{4.72}$$

c) Calculation of the coefficients $A_{oj}^{(2,1)}$

$$A_{oj}^{(2,1)} = (\lambda + \mu) \int_\Omega \frac{\partial \varphi_o}{\partial x_1} \cdot \frac{\partial \varphi_j}{\partial x_2} - \mu \int_0^L \varphi_o \frac{\partial \varphi_j}{\partial x_1} \, dx_1 . \tag{4.73}$$

Given that characteristic basis functions of nodes strictly interior at Ω are being considered, the border integral taken between $x_1 = 0$ and $x_1 = L$ in the expression (4.73) is identically nil.

In this case, the coefficients $A_{oj}^{(2,1)}$ are expressed as:

$$A_{oj}^{(2,1)} = (\lambda + \mu) \int_\Omega \frac{\partial \varphi_o}{\partial x_1} \cdot \frac{\partial \varphi_j}{\partial x_2} . \tag{4.74}$$

d) Calculation of the coefficient $A_{oj}^{(2,1)} (j = 0 \text{ to } 6)$.

By proceeding in a manner analogous to the evaluation of the coefficients $A_{oj}^{(1,1)}$, the following is obtained:

$$A_{oj}^{(2,1)} = (\lambda + \mu) , \tag{4.75}$$

$$A_{o1}^{(2,1)} = A_{o2}^{(2,1)} = A_{o4}^{(2,1)} = A_{o5}^{(2,1)} = -\frac{1}{2}(\lambda + \mu) , \tag{4.76}$$

$$A_{o3}^{(2,1)} = A_{o6}^{(2,1)} = \frac{1}{2}(\lambda + \mu) . \tag{4.77}$$

▶ **Estimation of the Second Member $b_0^{(1)}$**

In cases of a characteristic function of a node strictly interior at Ω, the second member (4.59) is expressed as:

$$b_o^{(1)} = \int_\Omega f_1 \varphi_o . \tag{4.78}$$

By using the trapezium quadrature formula, the following is obtained:

$$b_o^{(1)} \simeq 6 \times \frac{h^2}{2} \times \frac{1}{3} \times f_1^{(0)} = h^2 f_1^{(0)} . \tag{4.79}$$

A.11) The nodal equation corresponding to a characteristic basis function of a node strictly interior at Ω, is obtained by simple consolidation of the previous calculations.

It is then showed that it is expressed as:

$$+\mu\left[4\tilde{u}_1^{(0)}-\tilde{u}_1^{(1)}-\tilde{u}_1^{(2)}-\tilde{u}_1^{(4)}-\tilde{u}_1^{(5)}\right]+2(\lambda+\mu)\tilde{u}_1^{(0)}$$
$$-(\lambda+\mu)\left[\tilde{u}_1^{(1)}+\tilde{u}_1^{(4)}\right]+(\lambda+\mu)\tilde{u}_2^{(0)}$$
$$-\frac{(\lambda+\mu)}{2}\left[\tilde{u}_2^{(1)}+\tilde{u}_2^{(2)}-\tilde{u}_2^{(3)}+\tilde{u}_2^{(4)}+\tilde{u}_2^{(5)}-\tilde{u}_2^{(6)}\right]=h^2f_1^{(0)}.\qquad(4.80)$$

A.12) Navier's equation projected on the axis $(0;x_1)$ is written as:

$$(\lambda+2\mu)\frac{\partial^2 u_1}{\partial x_1^2}+(\lambda+\mu)\frac{\partial^2 u_2}{\partial x_1 \partial x_2}+\mu\frac{\partial^2 u_1}{\partial x_2^2}+f_1=0.\qquad(4.81)$$

In order to obtain a second order approximation of a second order crossed partial derivative, the Taylor's developments of a "sufficiently regular" solution u from equation (4.81) are written between the points $(x_1-h,\,x_2+h)$ and $(x_1,\,x_2)$, on one hand, and between the points $(x_1+h,\,x_2-h)$ and $(x_1,\,x_2)$, on the other.

This choice is guided by two reasons:

Since a second order approximation is sought, the developments have to be written up to the fourth order. From then on, the choice of points

$$M_{-h,+h}\equiv(x_1-h,x_2+h)\,,\quad M_{h,-h}\equiv(x_1+h,x_2-h)\quad\text{and}\quad M_{o,o}\equiv(x_1,x_2)$$

will enable playing along Taylor's development's symmetry to the appearing odd orders, particularly the third order, which has to be eliminated.

Moreover, in order to simplify writings, the notation convention will be as follows: $u_2\equiv u_2(M_{o,o})$.

$$u_2(M_{-h,+h})=u_2-h\frac{\partial u_2}{\partial x_1}+h\frac{\partial u_2}{\partial x_2}+\frac{h^2}{2}\frac{\partial^2 u_2}{\partial x_1^2}+\frac{h^2}{2}\frac{\partial^2 u_2}{\partial x_2^2}-h^2\frac{\partial^2 u_2}{\partial x_1 \partial x_2}$$
$$+\frac{1}{3!}\left[-h^3\frac{\partial^3 u_2}{\partial x_1^3}+3h^3\frac{\partial^3 u_2}{\partial x_1^2 \partial x_2}-3h^3\frac{\partial^3 u_2}{\partial x_1 x_2^2}+h^3\frac{\partial^3 u_2}{\partial x_2^3}\right]+O(h^4).$$
$$\qquad(4.82)$$

Similarly,

$$u_2(M_{h,-h})=u_2+h\frac{\partial u_2}{\partial x_1}-h\frac{\partial u_2}{\partial x_2}+\frac{h^2}{2}\frac{\partial^2 u_2}{\partial x_1^2}+\frac{h^2}{2}\frac{\partial^2 u_2}{\partial x_2^2}-h^2\frac{\partial^2 u_2}{\partial x_1 \partial x_2}$$
$$+\frac{1}{3!}\left[+h^3\frac{\partial^3 u_2}{\partial x_1^3}-3h^3\frac{\partial^3 u_2}{\partial x_1^2 \partial x_2}+3h^3\frac{\partial^3 u_2}{\partial x_1 x_2^2}-h^3\frac{\partial^3 u_2}{\partial x_2^3}\right]+O(h^4).$$
$$\qquad(4.83)$$

Adding equations (4.82) and (4.83) gives:

$$u_2(M_{-h,+h})+u_2(M_{h,-h})=2u_2+h^2\Delta u_2-2h^2\frac{\partial^2 u_2}{\partial x_1 \partial x_2}+O(h^4).\qquad(4.84)$$

The Laplacian Δu_2 at point $M_{o,o}$ is then replaced by its evaluation by the intervention of the centred finite-differences of the second order to yield the expression of the second order crossed partial derivative of u_2 at point $M_{o,o}$:

$$\frac{\partial^2 u_2}{\partial x_1 \partial x_2}(M_{o,o}) = \frac{u_2(M_{h,0}) + u_2(M_{-h,0}) + u_2(M_{0,h}) + u_2(M_{0,-h})}{2h^2} + \dots$$

$$\dots - \frac{2u_2(M_{0,0}) + u_2(M_{-h,+h}) + u_2(M_{h,-h})}{2h^2} + O(h^2) . \quad (4.85)$$

An approximation of the crossed partial derivative u_2 of the second order is obtained by omitting $O(h^2)$ from the equation (4.85) and by substituting the sequence of approximations to the true values of the solution, at the corresponding points following the notations used in the previous questions:

$$\frac{\partial^2 u_2}{\partial x_1 \partial x_2}(M_{o,o}) \simeq \frac{\tilde{u}_2^{(1)} + \tilde{u}_2^{(4)} + \tilde{u}_2^{(2)} + \tilde{u}_2^{(5)}}{2h^2} - \frac{2\tilde{u}_2^{(0)} + \tilde{u}_2^{(3)} + \tilde{u}_2^{(6)}}{2h^2} . \quad (4.86)$$

An approximation by finite-differences of equation (4.81) is then obtained by substitution of the second order crossed partial derivative by approximation (4.86) as well as other second order partial derivatives by their classical approximations.

The following is obtained:

$$(\lambda + 2\mu) \left[\frac{\tilde{u}_1^{(1)} - 2\tilde{u}_1^{(0)} + \tilde{u}_1^{(4)}}{h^2} \right] + \mu \left[\frac{\tilde{u}_1^{(2)} - 2\tilde{u}_1^{(0)} + \tilde{u}_1^{(5)}}{h^2} \right]$$

$$+ (\lambda + \mu) \left[\frac{\tilde{u}_2^{(1)} + \tilde{u}_2^{(4)} + \tilde{u}_2^{(2)} + \tilde{u}_2^{(5)} - 2\tilde{u}_2^{(0)} - \tilde{u}_2^{(3)} - \tilde{u}_2^{(6)}}{2h^2} \right] + f_1^{(0)} = 0 . \quad (4.87)$$

An ultimate re-organization of equation (4.87) yields the nodal equation (4.80) corresponding to a characteristic basis function of a node strictly interior at Ω.

Finally, the fact that partial derivatives, of the order $O(h^2)$, have been ignored in all approximations, the finite differences scheme is globally of the second order 2.

▶ **Basis Function φ_i Characteristic of a Node Interior at Σ_2**

A.13) Given that the border Σ_2 corresponds to the segment parametered by the equation $x_2 = 0$, the support of a characteristic basis function of a node of the mesh

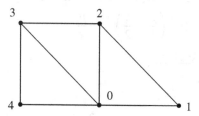

Fig. 4.6 Local Numeration Associated to a Node Strictly Interior at Σ_2.

belonging to this segment is composed of (according to local numeration, see below) T_{012}, T_{023} and T_{034} triangles.

Thus, calculation of the coefficients $A_{oj}^{(1,1)}$ is done by adapting the results of question **10** to the basis functions support φ_i mentioned above.

a) Calculation of the coefficients $A_{oj}^{(1,1)}$

$$A_{oo}^{(1,1)} = 2\mu + (\lambda + \mu) \, . \tag{4.88}$$

$$A_{o1}^{(1,1)} = -\frac{\mu}{2} - \frac{1}{2}(\lambda + \mu) \, , \tag{4.89}$$

$$A_{o2}^{(1,1)} = -\mu \, , \tag{4.90}$$

$$A_{o3}^{(1,1)} = 0 \, , \tag{4.91}$$

$$A_{o4}^{(1,1)} = A_{o1}^{(1,1)} \, . \tag{4.92}$$

b) Calculation of the coefficients $A_{oj}^{(2,1)}$

Expression of coefficients $A_{oj}^{(2,1)}$ is modified in relation to the one used in question **10**, considering the presence of the border integral of Σ_2.

In fact, the following is obtained:

$$A_{oj}^{(2,1)} = (\lambda + \mu) \int_\Omega \frac{\partial \varphi_o}{\partial x_1} \cdot \frac{\partial \varphi_j}{\partial x_2} - \mu \int_0^L \varphi_o \frac{\partial \varphi_j}{\partial x_1} \, dx_1 \, , \tag{4.93}$$

or even,

$$A_{oj}^{(2,1)} = (\lambda + \mu) \int_\Omega \frac{\partial \varphi_o}{\partial x_1} \cdot \frac{\partial \varphi_j}{\partial x_2} - \mu \int_{40} \varphi_o \frac{\partial \varphi_j}{\partial x_1} \, dx_1 - \mu \int_{01} \varphi_o \frac{\partial \varphi_j}{\partial x_1} \, dx_1 \, , \tag{4.94}$$

where it was noted symbolically as $\int_{40}$, the integral bearing an effect on the segment determined by the nodes 4 and 0, and, according to a similar convention, on the integral $\int_{01}$.

By using the trapezium quadrature formula for both one-dimensional integrals, it becomes:

$$A_{oo}^{(2,1)} = \frac{(\lambda + \mu)}{2} - \left(\frac{\mu}{h} \times \frac{h}{2} \right) + \left(\frac{\mu}{h} \times \frac{h}{2} \right) = \frac{(\lambda + \mu)}{2} \, , \tag{4.95}$$

$$A_{o1}^{(2,1)} = 0 - 0 - \left(\frac{\mu}{h} \times \frac{h}{2} \right) = -\frac{\mu}{2} \, , \tag{4.96}$$

$$A_{o2}^{(2,1)} = -\frac{(\lambda + \mu)}{2} \, , \tag{4.97}$$

$$A_{o3}^{(2,1)} = \frac{(\lambda + \mu)}{2} \, , \tag{4.98}$$

$$A_{o4}^{(2,1)} = -\frac{(\lambda + \mu)}{2} - \mu \int_{40} \varphi_o \frac{\partial \varphi_4}{\partial x_1} \, dx_1 = -\frac{(\lambda + \mu)}{2} + \frac{\mu}{2} \, . \tag{4.99}$$

▶ Estimation of the Second Member $b_0^{(1)}$

Again, calculation of the second member $b_0^{(1)}$ has to consider border integral of Σ_2, that is, interval $[0,L]$:

$$b_0^{(1)} = \int_\Omega f_1 \varphi_o + \int_0^L F_o \varphi_o . \qquad (4.100)$$

A numerical quadrature combining simultaneously the trapezium formula by triangle T_{ijk} and by segment $[x_i, x_{i+1}]$, leads to the following approximation:

$$b_0^{(1)} \simeq \frac{h^2}{2} f_1^{(0)} + h F_1^{(o)} . \qquad (4.101)$$

A.14) The nodal equation corresponding to a characteristic basis function of a mesh node coinciding with the border Σ_2 is obtained by grouping the estimations (4.87)–(4.92), (4.96)–(4.99) and (4.101):

$$(\lambda + 3\mu)\tilde{u}_1^0 - \left[\frac{\lambda}{2} + \mu\right](\tilde{u}_1^1 + \tilde{u}_1^4) - \mu\tilde{u}_1^2 + \frac{(\lambda + \mu)}{2}\tilde{u}_2^0$$

$$-\frac{\mu}{2}\tilde{u}_2^1 + \frac{(\lambda + \mu)}{2}[\tilde{u}_2^3 - \tilde{u}_2^2] - \frac{\lambda}{2}\tilde{u}_2^4 = \frac{h^2}{2}f_1^{(0)} + hF_1^{(o)} . \qquad (4.102)$$

A.15) In order to find the nodal equation (4.102) by the finite differences method, the projection $\sigma(-\mathbf{x}_2) \cdot \mathbf{x}_1$ of the boundary conditions on Σ_2 (4.14) has to be considered that bears an effect on the field of stress σ :

$$\frac{\partial u_1}{\partial x_2}(x_1, 0) + \frac{\partial u_2}{\partial x_1}(x_1, 0) = -\frac{F_1}{\mu} . \qquad (4.103)$$

Taylor's developments are then used in order to replace the combination of partial derivatives (4.103) by finite differences.

To achieve this, the component u_1 is measured at point $(x_1, x_2 + h)$ and the component u_2 at point $(x_1 + h, x_2)$, relative to the central point (x_1, x_2).

As for the developments shown in the answer to question **12**, the following notations will be used in order to simplify writings:

$$u_1 \equiv u_1(M_{0,0}), u_2 \equiv u_2(M_{0,0}), M_{0,+h} \equiv (x_1, x_2 + h) \text{ and } M_{+h,0} \equiv (x_1 + h, x_2)$$

$$u_1(M_{0,+h}) = u_1 + h\frac{\partial u_1}{\partial x_2} + \frac{h^2}{2}\frac{\partial^2 u_1}{\partial x_2^2} + O(h^3) , \qquad (4.104)$$

$$u_2(M_{+h,0}) = u_2 + h\frac{\partial u_2}{\partial x_1} + \frac{h^2}{2}\frac{\partial^2 u_2}{\partial x_1^2} + O(h^3) . \qquad (4.105)$$

The following is then obtain:

$$\frac{u_1(M_{0,+h}) + u_2(M_{+h,0}) - u_1(M_{o,o}) - u_2(M_{o,o})}{h}$$

$$= \left(\frac{\partial u_1}{\partial x_2} + \frac{\partial u_2}{\partial x_1}\right)(M_{o,o}) + \frac{h}{2}\frac{\partial^2 u_1}{\partial x_2^2}(M_{o,o}) + \frac{h}{2}\frac{\partial^2 u_2}{\partial x_1^2}(M_{o,o}) + O(h^2) . \qquad (4.106)$$

The first substitution now performed consists of replacing the sum of the first partial derivatives at point $(M_{o,o})$ by the second member of the equation of the boundary conditions (4.103).

$$\frac{u_1(M_{0,+h}) + u_2(M_{+h,0}) - u_1(M_{o,o}) - u_2(M_{o,o})}{h}$$

$$= -\frac{F_1(M_{o,o})}{\mu} + \frac{h}{2}\frac{\partial^2 u_1}{\partial x_2^2}(M_{o,o}) + \frac{h}{2}\frac{\partial^2 u_2}{\partial x_1^2}(M_{o,o}) + O(h^2) . \qquad (4.107)$$

The second substitution proposed allows an evaluation of the second partial derivative of

$$\frac{\partial^2 u_1}{\partial x_2^2}(M_{o,o})$$

in relation to the crossed partial derivative

$$\frac{\partial^2 u_2}{\partial x_1 \partial x_2}(M_{o,o})$$

by using the partial differential equation (4.81) whose approximation was shown by finite differences of the second order (4.86).

The following is thus obtained:

$$\frac{u_1(M_{0,+h}) + u_2(M_{+h,0}) - u_1(M_{o,o}) - u_2(M_{o,o})}{h} = \dots ,$$

$$-\frac{F_1(M_{o,o})}{\mu} - \frac{h}{2\mu}f_1(M_{o,o}) - \frac{(\lambda+2\mu)}{2\mu}h\frac{\partial^2 u_1}{\partial x_1^2}(M_{o,o}) + \dots ,$$

$$-\frac{(\lambda+2\mu)}{2\mu}h\frac{\partial^2 u_2}{\partial x_1 \partial x_2}(M_{o,o}) + \frac{h}{2}\frac{\partial^2 u_2}{\partial x_1^2}(M_{o,o}) + O(h^2) . \qquad (4.108)$$

By using notations relative to the local numeration, the scheme of the finite differences corresponding to the discretisation of the boundary conditions (4.103) is finally written as:

$$\mu\left[\tilde{u}_1^2 + \tilde{u}_2^1 - \tilde{u}_1^0 - \tilde{u}_2^0\right] = \dots ,$$

$$-hF_1^0 - \frac{h^2}{2}f_1^0 - \frac{(\lambda+2\mu)}{2}\left[\tilde{u}_1^1 - 2\tilde{u}_1^0 + \tilde{u}_1^4\right] + \dots ,$$

$$-\frac{\lambda+\mu}{2}\left[\tilde{u}_2^2 - \tilde{u}_2^3 - \tilde{u}_2^0 + \tilde{u}_2^4\right] + \frac{\mu}{2}\left[\tilde{u}_2^1 - 2\tilde{u}_2^0 + \tilde{u}_2^4\right] . \qquad (4.109)$$

It is then shown, after reorganising the equation (4.109), that the exact nodal equation (4.102) is found associated to a basis function, characteristic of a node interior at segment $[0, 1]$ of border Σ_2.

4.2 Clamped Plate

4.2.1 Statement

This problem seeks to study the two types of variational formulations for the vertical displacement equation u, following the axis $(O;\mathbf{z})$, of a square elastic plate Ω, which is perfectly clamped on its border and subjected to a density of efforts perpendicular to the plate, whose shape is: $\mathbf{f} = f(x,y)\mathbf{z}$, (cf. Fig. 4.7).

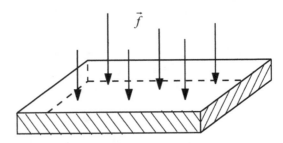

Fig. 4.7 Clamped Plate

▶ **Variational Formulations**

The interest is on the scalar function u of the variables (x,y), which is the solution to the 4th order partial derivatives of the equation:

Find $u \in H^4(\Omega)$ which is the solution to:

$$(\mathbf{CP_1}) \quad \begin{cases} \Delta^2 u = f \text{ in } \Omega, \\ u = \dfrac{\partial u}{\partial n} = 0 \text{ on } \partial\Omega, \end{cases} \tag{4.110}$$

where the bi-Laplacian operator Δ^2 is defined by:

$$\Delta^2 = \Delta(\Delta) = \frac{\partial^4}{\partial x^4} + 2\frac{\partial^4}{\partial x^2 \partial y^2} + \frac{\partial^4}{\partial y^4}, \tag{4.111}$$

Ω denotes the square $]0,1[\times]0,1[$ of the noted external normal $\mathbf{n}$, f is a given function belonging to $L^2(\Omega)$ and $H^4(\Omega)$ the functional space defined by:

$$H^4(\Omega) = \left\{ v: \Omega \subset \mathbf{R}^n \to \mathbf{R}, \frac{\partial^k v}{\partial x_{i_1} \dots \partial x_{i_k}} \in L^2(\Omega), \forall k = 0 \text{ to } 4 \right\}. \tag{4.112}$$

▶ **First Variational Formulation**

1) Let v be a test function "sufficiently regular" of the variables (x,y), show that a variational formulation associated to the continuous problem **(CP)** can be expressed as:

$$(\mathbf{VP}) \begin{cases} \text{Find } u \text{ belonging to } V \text{ solution to:} \\ \displaystyle\int_\Omega \Delta u \Delta v \, d\Omega = \int_\Omega f v \, d\Omega , \quad \forall v \in V . \end{cases} \tag{4.113}$$

The space V has to be specified (regularities and boundary conditions of functions v in V).

2) Can the finite elements $\mathbf{P}_1$ be used to solve in an approximate form the variational formulation **(VP)**?

▶ **Second Variational Formulation**

3) Let u be a solution of the problem $(\mathbf{CP}_1)$, the function φ is now introduced and defined by:

$$-\Delta u = \varphi . \tag{4.114}$$

If u is solution of the continuous problem $(\mathbf{CP}_1)$, show that the pair (φ, u) is a solution to the problem $(\mathbf{CP}_2)$:

Find $(\varphi, u) \in H^2(\Omega) \times H^2(\Omega)$, the solution to:

$$(\mathbf{CP}_2) \begin{cases} -\Delta\varphi = f \text{ in } \Omega , \\ -\Delta u \ = \varphi \text{ in } \Omega , \\ u = \dfrac{\partial u}{\partial n} = 0 \text{ on } \partial\Omega . \end{cases} \tag{4.115}$$

where:

$$H^2(\Omega) \equiv \left\{ v: \Omega \to \mathbf{R}, \ v \in L^2(\Omega) , \ \frac{\partial v}{\partial x_i} \in L^2(\Omega) , \ \frac{\partial^2 v}{\partial x_i x_j} \in L^2(\Omega) \right\} .$$

4) A double variational formulation is now performed by introducing a pair of test functions (ψ, v) belonging to $H_0^1(\Omega) \times H^1(\Omega)$, where $H_0^1(\Omega)$ is Sobolev's space defined by:

$$H_0^1(\Omega) = H^1(\Omega) \cap \{v/v = 0 \text{ on } \partial\Omega\} . \tag{4.116}$$

Show that if (φ, u) is a solution to the continuous problem $(\mathbf{CP_2})$ then (φ, u) is a solution to the following variational problem $(\mathbf{VP_2})$:

$$(\mathbf{VP_2}) \begin{cases} \text{Find } (\varphi, u) \text{ belonging to } H^1(\Omega) \times H_0^1(\Omega) \text{ solution to:} \\ a(\varphi, \psi) = L_f(\psi), \quad \forall \psi \in H_0^1(\Omega), \\ a(u, v) = L_\varphi(v), \quad \forall v \in H^1(\Omega), \end{cases} \qquad (4.117)$$

where the bilinear form $a(.,.)$ as well as the linear form $L_f(.)$ will be determined.

It is to be noted that the linear form $L_\varphi(.)$ is identical to $L_f(.)$, as soon as f is substituted by φ.

▶ **Approximation by Finite Elements P_1**

An approximation of the variational problem $(\mathbf{VP_2})$ is now performed by finite elements $\mathbf{P_1}$.

To achieve this, a constant step of discretization h is introduced and the square Ω is uniformly meshed by triangles $T_k, (k = 1 \text{ to } N_{\text{Triangles}})$, isoceles rectangles of side h, (cf. Fig. 4.8).

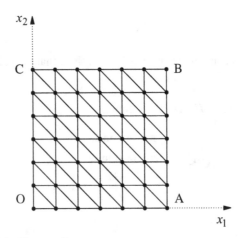

Fig. 4.8 Mesh by Finite Elements P_1

The sequence of $M_{i,j}$ points would have thus been constructed with (x_i, y_j) coordinates defined by:

$$\begin{cases} x_0 = y_0 = 0, \quad x_{N+1} = y_{N+1} = 1, \\ x_{i+1} = x_i + h, \quad i = 1 \text{ to } N+1, \\ y_{j+1} = y_j + h, \quad j = 1 \text{ to } N+1. \end{cases} \qquad (4.118)$$

5) Let $\tilde{V}$ be the space defined by:

$$\tilde{V} = \{ \tilde{\psi} : \Omega \longrightarrow \mathbf{R}, \tilde{\psi} \in C^o(\Omega), \tilde{\psi}|_{T_k} \in P_1(T_k), k = 1 \text{ to } N_{\text{Triangles}} \}, \qquad (4.119)$$

where $P_1(T_k)$ represents the set of polynomial having a degree less or equal to 1 in relation to the ordered pair (x, y).

Moreover it is to be noted that $(\xi_i), (i = 1$ to $(N+2)^2)$ the canonical basis of the space $\tilde{V}$, that is it satisfies the property: $\xi_l(M_m) = \delta_{lm}$.

On the other hand, $\tilde{V}_0$ will refer to the space of functions belonging to $\tilde{V}$, being zero on the border of Ω.

The formula is stated as follows:

$$\tilde{\varphi} = \sum_{j=1,(N+2)^2} \tilde{\varphi}_j \xi_j \text{ and } \tilde{u} = \sum_{j=1,N^2} \tilde{u}_j \xi_j . \qquad (4.120)$$

Show that the approximate variational formulation to the problem $(\mathbf{VP_2})$ is written as:

$$(\widetilde{\mathbf{VP_2}}) \begin{cases} \displaystyle\sum_{j=1,(N+2)^2} A_{ij}\varphi_j = b_i^1 , & (i = 1 \text{ to } N^2) , \\[2mm] \displaystyle\sum_{j=1,N^2} A_{ij}\tilde{u}_j = b_i^2 , & (i = 1 \text{ to } (N+2)^2) , \\[2mm] \text{with: } A_{ij} = \displaystyle\int_\Omega \nabla \xi_i \cdot \nabla \xi_j \, d\Omega , \\[2mm] b_i^1 = \displaystyle\int_\Omega f \xi_i \, d\Omega , \quad b_i^2 = \displaystyle\int_\Omega \tilde{\varphi} \xi_i \, d\Omega . \end{cases} \qquad (4.121)$$

▶ **System of Equations Associated with $\tilde{\varphi}$**

6) Given the mesh regularity (invariance by horizontal and vertical translation), local numeration described is adopted as in Fig. 4.9. Then make explicit that part of system (4.121) concerning $\tilde{\varphi}$, written in its local form:

$$\sum_{j=0,6} A_{oj}\tilde{\varphi}_j = b_o^1 . \qquad (4.122)$$

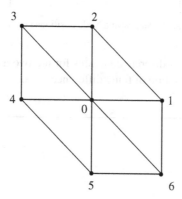

Fig. 4.9 Local Numeration Associated with a Node Strictly Interior at Ω

– The coefficients A_{oj} are then exactly calculated whereas the second member b_0^1 will be approximately evaluated by means of the trapezium quadrature method.

7) What is the essential characteristic of system (4.122) which penalizes the conventional implementation of a linear system inversion algorithm.

▶ **System of Equations Associated with $\tilde{u}$**

Nodal equation associated with a characteristic basis function of *a node interior at Ω.*

8) Wiser from the experience of the previous question, directly write the equations of system (4.121) corresponding to the characteristic basis functions of a node situated in the interior of the mesh.

Equation associated with *a characteristic basis function* of *a node interior to segment OA.*

9) Still by adopting a local numeration (cf. Fig. 4.10), make explicit the system of equations (4.121) written in its local form:

$$\sum_{j=2,3} A_{oj}\tilde{u}_j = b_o^2 . \tag{4.123}$$

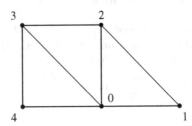

Fig. 4.10 Local Numeration Associated with a Node Positioned on the $\partial\Omega$ Border

10) Directly find the set of discrete equations for the two approximation functions $(\tilde{\varphi}, \tilde{u})$, by using a discretisation of finite differences type.

4.2.2 Solution

▶ **First Variational Formulation**

A test function v is introduced, defined on Ω having real values, and the partial differential equation of the continuous problem $(\mathbf{CP_1})$ is multiplied by v:

$$\int_\Omega \Delta^2 u \cdot v \, d\Omega = \int_\Omega f \cdot v \, d\Omega . \tag{4.124}$$

Green's formula is used twice successively, the left member of (4.124) is written as follows:

$$\int_\Omega \Delta^2 u \cdot v \, d\Omega = \int_\Omega \Delta(\Delta u) \cdot v \, d\Omega = -\int_\Omega \nabla(\Delta u) \cdot \nabla v \, d\Omega + \int_{\partial\Omega} \frac{\partial(\Delta u)}{\partial n} \cdot v \, d\Gamma ,$$

$$= \int_\Omega \Delta u \cdot \Delta v \, d\Omega - \int_{\partial\Omega} \Delta u \cdot \frac{\partial v}{\partial n} \, d\Gamma + \int_{\partial\Omega} \frac{\partial(\Delta u)}{\partial n} \cdot v \, d\Omega . \tag{4.125}$$

Now and as presented systematically without any *ex ante* concerning the nature and properties of test functions v, the construction of the functional frame V corresponding to the frame of existence of functions v, should meet the requirements for conserving the set of information existing in the formulation of the continuous problem $(\mathbf{CP_1})$.

However, it turns out that none of the boundary conditions leading to the clamping of the plate along its border $\partial\Omega$ appears in the integral formulation (4.125).

That is the reason why functions v of V are required to meet the following boundary conditions:

$$v = \frac{\partial n}{\partial n} = 0, \text{ on } \partial\Omega . \tag{4.126}$$

Thus, as the object of the search for a solution to the variational formulation consists in finding solution u in space V, it will be ascertained that this solution will keep the boundary conditions to which it is subjected as soon as the existence of such a solution has been proved.

By taking into account the boundary conditions (4.126), the last equation of (4.125) is written:

$$\int_\Omega \Delta^2 u \cdot v \, d\Omega = \int_\Omega \Delta u \cdot \Delta v \, d\Omega . \tag{4.127}$$

Then (4.127) is replaced in (4.124) and the variational equation is obtained:

$$\text{Find } u \in V \text{ the solution to: } \int_\Omega \Delta u \cdot \Delta v \, d\Omega = \int_\Omega f \cdot v \, d\Omega \tag{4.128}$$

The final stage of variational formulation consists in defining the functional space V.

To make this happen, only regularity properties have to be added to boundary conditions in (4.126) to guarantee the integral convergence of equation (4.128).

The integral on the left side of equation (4.128) can be controlled by:

$$\left| \int_\Omega \Delta u \cdot \Delta v \, d\Omega \right| \leq \int_\Omega |\Delta u \cdot \Delta v| \, d\Omega \leq \left[\int_\Omega (\Delta u)^2 \, d\Omega \right]^{1/2} \cdot \left[\int_\Omega (\Delta v)^2 \, d\Omega \right]^{1/2},$$

(4.129)

where the Cauchy-Schwartz inequality would have been used.

Then consider the natural norm of the Sobolev space $H^2(\Omega)$:

$$\forall u \in H^2(\Omega):$$

$$\|u\|_{H^2(\Omega)}^2 = \left[\|u\|_{L^2(\Omega)}^2 + \sum_{i=1,2} \left\| \frac{\partial u}{\partial x_i} \right\|_{L^2(\Omega)}^2 + \sum_{i,j=1,2} \left\| \frac{\partial^2 u}{\partial x_i \partial x_j} \right\|_{L^2(\Omega)}^2 \right],$$

(4.130)

with:

$$\forall u \in L^2(\Omega): \|u\|_{L^2(\Omega)}^2 = \int_\Omega |u|^2 \, d\Omega.$$

(4.131)

Consequently, the following is obtained:

$$\forall u \in H^2(\Omega): \left[\int_\Omega (\Delta u)^2 \, d\Omega \right] \leq \|u\|_{H^2(\Omega)}^2.$$

(4.132)

Then the integral of the member on the left side of (4.128) is used again by means of the inequality (4.129):

$$\left| \int_\Omega \Delta u \cdot \Delta v \, d\Omega \right| \leq \|u\|_{H^2(\Omega)} \|v\|_{H^2(\Omega)}.$$

(4.133)

In other words, considering u and v in $H^2(\Omega)$ is enough to guarantee the convergence of the integral in the left member of the variational equation (4.128).

Likewise, by using once again the Cauchy-Schwartz inequality, the member on the right side of equation (4.128) is increased by:

$$\left| \int_\Omega f \cdot v \, d\Omega \right| \leq \int_\Omega |f \cdot v| \, d\Omega \leq \|f\|_{L^2(\Omega)} \|u\|_{L^2(\Omega)}.$$

(4.134)

Then introduce the Sobolev space $H_0^2(\Omega)$ defined by:

$$H_0^2(\Omega) = H^2(\Omega) \cap \left\{ v / v = \frac{\partial v}{\partial n} = 0, \text{ on } \partial\Omega \right\}.$$

(4.135)

Finally, the variational problem **(VP)** is written as:

$$\textbf{(PV)} \int_{\Omega} \Delta u \Delta v \, d\Omega = \int_{\Omega} f v \, d\Omega , \quad \forall v \in H_0^2(\Omega) . \qquad (4.136)$$

A.2) The use of finite elements P_1 to solve the variational formulation **(VP)** by approximation would lead to the fact that the member on the left side of (4.136) would be identically nil.

In fact, the second partial derivatives of a function whose degree is less or equal to one with respect to the ordered pair (x,y) being nil, it follows that the Laplacian of such a function is alike!

Thus, the use of such finite elements to numerically solve the problem **(VP)** by variational approximation is not really recommended since the right side of (4.136) is not zero.

▶ **Second Variational Formulation**

A.3) The formulation of problem $(\textbf{CP}_2)$ is instantaneous. In fact, it is only necessary to inject of the function change (4.114) in the partial differential equation of the plates of problem $(\textbf{CP}_1)$ to obtain the first partial differential equation to problem $(\textbf{CP}_2)$, i. e. Laplace equation:

$$-\Delta \varphi = f .$$

The other equations forming the rest of problem $(\textbf{CP}_2)$ are trivial.

A.4) Let (ψ, v) be a couple of test functions belonging to $H_0^1(\Omega) \times H^1(\Omega)$. Function ψ will be the test function associated with function φ and function v with the second unknown u.

In order to obtain a variational formulation $(\textbf{VP}_2)$ associated with the continuous problem $(\textbf{CP}_2)$, multiply each of the partial differential equations of continuous problem $(\textbf{CP}_2)$ by its corresponding test function and obtain:

$$\textbf{(PV}_2\textbf{)} \begin{cases} -\int_{\Omega} \Delta \varphi \cdot \psi \, d\Omega = \int_{\Omega} \nabla \varphi \cdot \nabla \psi \, d\Omega + \int_{\partial \Omega} \frac{\partial \varphi}{\partial n} \cdot \psi \equiv \int_{\Omega} f \cdot \psi \, d\Omega , \\ -\int_{\Omega} \Delta u \cdot v \, d\Omega = \int_{\Omega} \nabla u \cdot \nabla v \, d\Omega + \int_{\partial \Omega} \frac{\partial u}{\partial n} \cdot v \equiv \int_{\Omega} \varphi \cdot v \, d\Omega . \end{cases} \qquad (4.137)$$

Then use the boundary conditions bearing upon u and its normal differential coefficient, identically nil on the border $\partial \Omega$ of Ω.

In fact, concerning the normal differential coefficient of u, given the fact that the values of u do not intervene at all in the integrals of double formulation (4.137), in order to keep its memory in variational formulation $(\textbf{VP}_2)$, introduce the suitable properties in the functional space in which function ψ will be the generic element.

It really consists of the variational space associated with ψ since the plate equation is multiplied by this function, of course rewritten with a change of function φ.

Thus, when the boundary conditions bearing on the nullity of u on the border $\partial\Omega$ is replaced in system (4.137) and by choosing $H^1(\Omega) \times H_0^1(\Omega)$ as functional spaces for the unknowns (φ, u) on one hand, and $H_0^1(\Omega) \times H^1(\Omega)$ for the pair of test functions (ψ, v) on the other hand, the double variational formulation is written as:

Find (φ, u) belonging to $H^1(\Omega) \times H_0^1(\Omega)$ being the solution to:

$$(\mathbf{PV_2}) \begin{cases} a(\varphi, \psi) \equiv \int_\Omega \boldsymbol{\nabla}\varphi \cdot \boldsymbol{\nabla}\psi \, d\Omega = \int_\Omega f \cdot \psi \, d\Omega \equiv L_f(\psi) , & \forall \psi \in H_0^1(\Omega) , \\ a(u, v) \equiv \int_\Omega \boldsymbol{\nabla}u \cdot \boldsymbol{\nabla}v \, d\Omega = \int_\Omega \varphi \cdot v \, d\Omega \equiv L_\varphi(v) , & \forall v \in H^1(\Omega) . \end{cases}$$

$$(4.138)$$

A.5) The change from variational formulation $(\mathbf{VP_2})$ to approximate variational formulation $\widetilde{(\mathbf{VP_2})}$ is obtained by substituting unknowns (φ, u) by respective approximation functions $(\tilde{\varphi}, \tilde{u})$:

Find $(\tilde{\varphi}, \tilde{u})$ belonging to $\tilde{V} \times \tilde{V}_0$, being solution to:

$$\widetilde{(\mathbf{VP_2})} \begin{cases} a(\tilde{\varphi}, \tilde{\psi}) = L_f(\tilde{\psi}) , & \forall \tilde{\psi} \in H_0^1(\Omega) , \\ a(\tilde{u}, \tilde{v}) = L_\varphi(\tilde{v}) , & \forall \tilde{v} \in H^1(\Omega) . \end{cases}$$

$$(4.139)$$

Then the decompositions of $\tilde{\varphi}$ and $\tilde{u}$ are used on their respective basis, according to the formula (4.120).

Moreover, given the bilinearity of form $a(.,.)$ and the linearity of form $L_f(.)$, (form $L_\varphi(.)$ obviously displays the same property), the expressions of (4.121) easily follows it.

One will note on this occasion that approximation $\tilde{u}$ breaks up on basis functions N^2 where as approximation $\tilde{\varphi}$ breaks up on basis functions $(N+2)^2$.

In fact, this is due to the fact that unknown $\tilde{u}$ has to be identically nil on the border $\partial\Omega$.

As a result, the characteristic basis functions of nodes situated on the border of square Ω should not be included in the decomposition of function $\tilde{u}$.

The result is that only the points N^2 strictly interior to the mesh and the characteristic basis functions of these nodes have to be considered in the decomposition of approximation $\tilde{u}$.

The same arguments lead to N^2 equations whose second member is mentioned as b_i^1 in the variational problem formulation and $(N+2)^2$ equations for the second system whose second member is mentioned as b_i^2.

A.6) Now the constitution of the linear system associated with the unknown $\tilde{\varphi}$ of problem $(\mathbf{VP_2})$ is considered as written in its local form.

To make that happen, it is observed that having considered a regular mesh, (cf. Fig. 4.8), and by adopting the locale numeration shown in Fig. 4.9, if 0 becomes the local number of numbered function i in the global numeration, only the numbers of the triangles' apexes, where the node 0 is one of the apexes, intervene, and this, in accordance with the properties of the "affine by triangle" basis function supports.

Then the following is obtained:

$$\left[\sum_{j=1,(N+2)^2} A_{ij}\varphi_j = b_i^1 \ , \ i=1 \text{ to } N^2\right] \Rightarrow \left[\sum_{j=0,6} A_{oj}\tilde{\varphi}_j = b_o^1\right] \tag{4.140}$$

▶ **Calculation of Coefficients A_{00}**

All the coefficients will be calculated correctly, given the fact that the gradients of basis functions ξ_i are constant per triangle.

These calculations are conventional and may be consulted, for further details, in the work of D. Euvrard, [4].

a) Calculation of coefficient A_{00}

$$A_{00} = \int_{\text{Supp } \xi_0} |\nabla\xi_0|^2 \, d\Omega = \int_{012} |\nabla\xi_0|^2 \, d\Omega + \int_{023} |\nabla\xi_0|^2 \, d\Omega$$

$$+ \int_{034} |\nabla\xi_0|^2 \, d\Omega + \int_{045} |\nabla\xi_0|^2 \, d\Omega + \int_{056} |\nabla\xi_0|^2 \, d\Omega + \int_{061} |\nabla\xi_0|^2 \, d\Omega \, ,$$

$$A_{00} = \frac{h^2}{2} \left\{ \frac{2}{h^2} + \frac{1}{h^2} + \frac{1}{h^2} + \frac{2}{h^2} + \frac{1}{h^2} + \frac{1}{h^2} \right\} = 4 \, . \tag{4.141}$$

b) Calculation of coefficients A_{01} and A_{02}

$$A_{01} = \int_{\text{Supp } \xi_0 \cap \text{Supp } \xi_1} \nabla\xi_0 \cdot \nabla\xi_1 \, d\Omega = \int_{012} \nabla\xi_0 \cdot \nabla\xi_1 \, d\Omega + \int_{061} \nabla\xi_0 \cdot \nabla\xi_1 \, d\Omega$$

$$= \frac{h^2}{2} \left\{ -\frac{1}{h^2} - \frac{1}{h^2} \right\} = -1 \, . \tag{4.142}$$

By analogue reasoning, the following would be found in the same way:

$$A_{02} = \int_{\text{Supp } \xi_0 \cap \text{Supp } \xi_2} \nabla\xi_0 \cdot \nabla\xi_2 \, d\Omega = \int_{012} \nabla\xi_0 \cdot \nabla\xi_2 \, d\Omega + \int_{023} \nabla\xi_0 \cdot \nabla\xi_2 \, d\Omega$$

$$= \frac{h^2}{2} \left\{ -\frac{1}{h^2} - \frac{1}{h^2} \right\} = -1 \, . \tag{4.143}$$

c) Calculation of coefficient A_{03}

$$A_{03} = \int_{\text{Supp } \xi_0 \, \cap \, \text{Supp } \xi_3} \nabla \xi_0 \cdot \nabla \xi_3 \, d\Omega = \int_{023} \nabla \xi_0 \cdot \nabla \xi_3 \, d\Omega + \int_{034} \nabla \xi_0 \cdot \nabla \xi_3 \, d\Omega$$

$$= \frac{h^2}{2} \{0 + 0\} = 0 \,. \tag{4.144}$$

d) Calculation of coefficients A_{04}, A_{05} and A_{06}

Because of symmetries inherent to the invariance of the mesh in the two directions of the plan, on one hand at the symmetry of bilinear form $a(.,.)$, and consequently of matrix A_{0j}, the following is obtained:

$$\begin{cases} A_{04} = A_{40} = A_{01} = -1 \,, \\ A_{05} = A_{50} = A_{02} = -1 \,, \\ A_{06} = A_{60} = A_{03} = 0 \,. \end{cases} \tag{4.145}$$

▶ **Estimation of the Second Member b_0^1**

Evaluation of the second member b_0^1 is performed using the trapezium quadrature formula:

$$\iint_T f(\chi) \, d\chi \simeq \frac{\text{Area } T}{3} \{ f(A_i) + f(A_j) + f(A_k) \} \,, \tag{4.146}$$

where T refers to any triangle of the mesh whose vertices A_i, A_j and A_k would have been noted. Thus,

$$b_0^1 = \int_{\text{Supp } \xi_0} f \xi_0 \, d\Omega = \int_{012} f \xi_0 \, d\Omega + \int_{023} f \xi_0 \, d\Omega + \int_{034} f \xi_0 \, d\Omega$$

$$+ \int_{045} f \xi_0 \, d\Omega + \int_{056} f \xi_0 \, d\Omega + \int_{061} f \xi_0 \, d\Omega \,,$$

$$\simeq \frac{h^2}{2} \times 6 \times \frac{1}{3} \times [f_0 \times 1] \,, \tag{4.147}$$

where f_0 refers to the value of the second member f at node 0.

The generic equation (4.140) associated with characteristic basis function ξ_0 of an interior node (i. e., equal to 1 in this node and to 0 at the other nodes of the mesh) and noted 0 in local numeration, is written as:

$$4 \tilde{\varphi}_0 - \tilde{\varphi}_1 - \tilde{\varphi}_4 - \tilde{\varphi}_2 - \tilde{\varphi}_5 = h^2 f_0 \,. \tag{4.148}$$

As usual, and for any regular mesh, the finite differences schema associated to the Laplacian is observed:

$$-\left[\frac{\tilde{\varphi}_1 - 2\tilde{\varphi}_0 + \tilde{\varphi}_4}{h^2}\right] - \left[\frac{\tilde{\varphi}_2 - 2\tilde{\varphi}_0 + \tilde{\varphi}_5}{h^2}\right] = f_0 . \qquad (4.149)$$

A.7) The linear system described in the generic equation (4.148) is a system of $(N+2)^2$ equations with N^2 unknowns.

In other words, it is a rectangular linear system. Thus, not having as many equations as unknowns, it cannot be solved numerically in an autonomous manner.

A complete resolution by simultaneously finding out the pair of unknowns $(\tilde{\varphi}, \tilde{u})$ may only be performed by completing this system with the ones dealt with in questions **8)** and **9)**.

But hold on since each thing is performed in its own time...

A.8) Given that the linear system governing the approximation function $\tilde{u}$ is built up with a same germ matrix, the basis function of a node strictly interior to Ω may be written as the following corresponding nodal equation:

$$4\tilde{u}_0 - \tilde{u}_1 - \tilde{u}_4 - \tilde{u}_2 - \tilde{u}_5 = h^2 \varphi_0 . \qquad (4.150)$$

However, the second member f_0 of system (4.148) should be interchanged with the appropriate value of φ_0 before carrying out the above operation.

A.9) Now a characteristic function of a node belonging to a segment OA is considered, one of the segments forming the border $\partial\Omega$, (cf. Fig 4.8).

In this case, only the functions of nodes locally numbered $0, 1, 2, 3$ and 4 display a potential contribution in the writing of the nodal equation associated with the characteristic function of node 0.

In other words, the nodal equation is in this case written as:

$$\sum_{j=0,4} A_{0j}\tilde{u}_j = \sum_{j=2,3} A_{0j}\tilde{u}_j = b_0^2 . \qquad (4.151)$$

It would be observed that the contributions of nodes 0, 1 and 4 are at once eliminated because they are situated on the border of Ω domain where the unknown u has to be identically nil. The same thing is done for the approximation function $\tilde{u}$.

The two coefficients A_{02} and A_{03} are thus basically obtained in accordance with the calculations shown in detail in question **6)**.

Thus, the following is immediately obtained:

$$A_{02} = \int_{\text{Supp } \xi_0 \cap \text{Supp } \xi_2} \nabla \xi_0 \cdot \nabla \xi_2 \, d\Omega = \int_{012} \nabla \xi_0 \cdot \nabla \xi_2 \, d\Omega + \int_{023} \nabla \xi_0 \cdot \nabla \xi_2 \, d\Omega ,$$

$$= -1 .$$

$$A_{03} = \int_{\text{Supp } \xi_0 \cap \text{Supp } \xi_3} \nabla \xi_0 \cdot \nabla \xi_3 \, d\Omega = \int_{023} \nabla \xi_0 \cdot \nabla \xi_3 \, d\Omega + \int_{034} \nabla \xi_0 \cdot \nabla \xi_3 \, d\Omega ,$$

$$= 0 . \tag{4.152}$$

As the same causes lead to the same effects, the second member b_0^2 is estimated by analogous manner as for the second member b_0^1, so long as it is considered that in the present case there are twice fewer triangles and that function $\tilde{\varphi}$ replaces function f in the integral of second member b_0^1.

Finally, the nodal equation is written as:

$$-\tilde{u}_2 = \frac{h^2}{2} \varphi_0 . \tag{4.153}$$

A.10) The discretisation by finite differences of problem $(\mathbf{CP_2})$ is conventional. It consists in discretising successively two second-order Laplacians and one Neumann condition.

The solution to Dirichlet's and Neumann's problems that have been presented on this subject may be consulted, (cf. Problems [3.1] and [3.2]), for the integrality of discretisation calculations by finite differences.

Chapter 5
Finite Elements Applied to Strength of Materials

Preamble

This chapter is dedicated to the application of the finite elements method within the framework of strengths of materials.

The main objective of this chapter is to clear up, as much as possible, a state of confusion that predominates within the community of graduate students in mechanics, physics, and also within certain graduate schools of engineering.

In fact, experience shows that, very often, the finite elements method appears to be radically different and this causes an unjustified confusion in students' minds! – depending on whether this method is presented within the framework of a classic course in numerical analysis or whether it relates to a course in solid mechanics.

The author's expressed wish and the essential motivation in writing this chapter are to get rid of any doubts that may exist among the concerned public by demonstrating the uniqueness in fundament and in form of the finite elements method, more so as it originated from the mechanics of deformable solids.

For, in fact, only the teacher's lack of concern is the unique source of possible confusions. It is the teaching staff's responsibility to dissipate the difficulty unnecessarily faced by a great number of students.

Let there not be the slightest ambiguity. The finite elements method though applied to solid mechanics is, and should be standardized for the benefit of students on one hand, and for its own further use in applications that can only benefit from its practical and indisputable performance and flexibility on the other hand.

In order to set up this standardization, the aim of this presentation is to propose and expound, within the framework of the beam theory, a double application of the finite elements method.

Therefore, on one hand, the "numerical analysis" version based on the approximation of a variational formulation will be developed, and, on the other hand, the "mechanics of deformable solids" version aimed at resolving the minimization problem associated with and equivalent to the variational formulation of the first version will be expanded.

Moreover, as far as the "solid mechanics version" of the finite elements method is concerned, this opportunity is taken to address the principles of the technique of assembly of the linear system obtained during the approximation of the minimization problem.

5.1 Beam Subjected to Simple Traction

5.1.1 Case of a Restraint

Statement

Consider a homogenous beam Ω having a length L, a cross section S and a constant density ρ, whose mechanical behaviour is the isotropic elasticity with small disturbances.

Moreover, the constituent material of the beam has a Young's modulus E and a Poisson's ratio v.

$(O; \mathbf{X}_1)$ refers to the axis of the beam and the current abscissa is x. The beam is restrained at the abscissa $x = 0$ and is free from stress when $x = L$.

In addition, a charge density $\mathbf{f} = f_1 \mathbf{X}_1$ is applied to longitudinal forces along the beam (see Fig. 5.1). The density f_1 is given and has "enough regularity" to enable the integration calculations of the theoretical part.

▶ **Displacement Variational Formulation – Principle of Virtual Works – Theoretical Part**

1) By choosing a field of virtual displacements $\mathbf{U}^*$ in the form:

$$\mathbf{U}^* = u^*(x)\mathbf{X}_1 \quad \text{such that} \quad u^*(x) = 0 \, , \tag{5.1}$$

find the virtual work of the interior forces T_{Int}^* and the virtual work of the exterior forces T_{Ext}^* defined by:

$$T_{\text{Int}}^* = -\int_\Omega \sigma_{ij}\varepsilon_{ij}(\mathbf{U}^*)\,\mathrm{d}\Omega, \quad T_{\text{Ext}}^* = \int_\Omega \mathbf{f} \cdot \mathbf{U}^*\,\mathrm{d}\Omega \, , \tag{5.2}$$

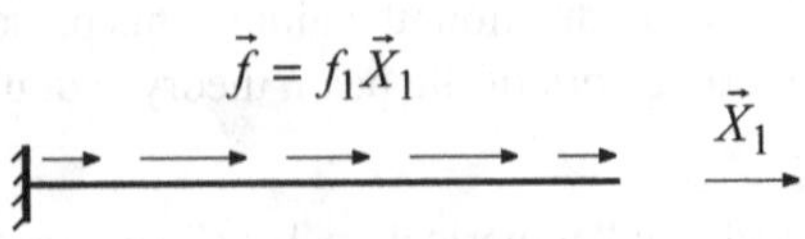

Fig. 5.1 Beam Subject to Traction

where σ refers to the stress tensor and $\varepsilon(\mathbf{U}^*)$ to the linear strain tensor associated with the virtual displacements field $\mathbf{U}^*$:

$$\varepsilon_{ij}(\mathbf{U}^*) = \frac{1}{2}\left[\frac{\partial U_i^*}{\partial x_j} + \frac{\partial U_j^*}{\partial x_i}\right]. \tag{5.3}$$

It would be necessary to introduce the normal force $N(x)$ and the force loading $f(x)$ defined by:

$$N(x) = \iint_{S(x)} \sigma_{11}\, dS(x), \quad f(x) = \iint_{S(x)} f_1\, dS(x), \tag{5.4}$$

$S(x)$ refers to the section having abscissa x.

2) Assuming that the various integrated magnitudes are "sufficiently regular", show that the application of the Principle of Virtual Works leads to the following formal variational formulation (**EVP**):

$$(\mathbf{EVP}) \quad \left[\begin{array}{l} \text{Find } N \text{ defined on } [0,L] \text{ having values in } \mathbf{R}, \text{ solution to:} \\[2mm] \displaystyle\int_0^L \dot{N}u^*\,dx + N(L)u^*(L) + \int_0^L fu^*\,dx = 0, \\[4mm] \forall u^*/u^*(0) = 0, \end{array}\right. \tag{5.5}$$

the parameters having been defined: $\dot{N} \equiv \dfrac{dN}{dx}$.

3) Using the behaviour law of the material ($N = ES\dot{u}$) and by assuming that the force density f belongs to $L^2(0,L)$, show that the continuous problem (**CP**) consisting of the equilibrium equations of the beam is written as:

$$(\mathbf{CP}) \quad \left[\begin{array}{l} \text{Find } u \text{ belonging to } H^2(0,L), \text{ solution to:} \\[2mm] -ES\ddot{u}(x) = f(x), \quad \forall x \in]0,L[, \\[2mm] u(0) = 0, \quad \dot{u}(L) = 0. \end{array}\right. \tag{5.6}$$

4) $V \equiv H_*^1(0,L)$ refers to the Sobolev space defined by:

$$H_*^1(0,L) = \{v \in L^2(0,L), v' \in L^2(0,L) \text{ such that } v(0) = 0\}, \tag{5.7}$$

then show that a displacement variational formulation (**VP**) can be written as:

$$(\mathbf{VP}) \quad \left[\begin{array}{l} \text{Find } u \text{ belonging to } V, \text{ solution to:} \\[2mm] \displaystyle ES\int_0^L \dot{u}(x)\dot{v}(x)\,dx = \int_0^L f(x)v(x)\,dx, \quad \forall v \in V. \end{array}\right. \tag{5.8}$$

5) Likewise, show that there is a minimization problem (**MP**) equivalent to the variational formulation (**VP**) defined by:

$$(\mathbf{MP}) \quad \left[\begin{array}{l} \text{Find } u \text{ belonging to } V, \text{ solution to:} \\[2mm] \displaystyle J(u) = \operatorname*{Min}_{v\in V} J(v), \\[4mm] \text{where: } J(v) = \dfrac{ES}{2}\int_0^L \dot{v}^2(x)\,dx - \int_0^L f(x)v(x)\,dx. \end{array}\right. \tag{5.9}$$

▶ **Numerical Part**

This part is dedicated to the approximation by finite elements applied to elastic beams.

In order to propose the application of this method as a complementary to the one exposed in the previous chapters, this approximation will be worked out by underlining the assembly technique on one hand and by applying it in the particular framework of approximation of the minimization problem (**MP**) on the other hand.

The global framework of the approximation is that of the finite elements P_1 and a regular mesh of the interval $[0, L]$ having constant step h is considered, such as:

$$\begin{cases} x_0 = 0, \quad x_{N+1} = L, \\ x_{i+1} = x_i + h, \quad i = 0 \text{ to } N. \end{cases} \tag{5.10}$$

The approximation space $\tilde{V}$ is now defined using:

$$\tilde{V} = \left\{ \tilde{v} \colon [0, L] \to \mathbf{R}, \ \tilde{v} \in C^0([0, L]), \ \tilde{v}|_{[x_i, x_{i+1}]} \in P_1([x_i, x_{i+1}]), \tilde{v}(0) = 0 \right\}, \tag{5.11}$$

where $P_1([x_i, x_{i+1}])$ refers to the space of polynomials defined on $[x_i, x_{i+1}]$, having a degree less than or equal to one.

6) If the canonical basis of $\tilde{V}$ satisfying $\varphi_i(x_j) = \delta_{ij}$, $(\delta_{ii} = 1$, and $\delta_{ij} = 0$ if $i \neq j)$, is noted $\varphi_i, (i = 1 \text{ to } N + 1)$, let $\tilde{v}$ belongs to $\tilde{V}$ defined by:

$$\tilde{v} = \sum_{j=1}^{N+1} \tilde{v}_j \varphi_j. \tag{5.12}$$

Propose a mechanical interpretation having coefficients $\tilde{v}_j$.

7) Let $(\widetilde{\text{MP}})$ be the approximate minimization problem associated with the problem (**MP**) defined by:

$$(\widetilde{\text{MP}}) \text{ Find } \tilde{u} \text{ belonging to } \tilde{V}, \text{ solution of: } J(\tilde{u}) = \underset{\tilde{v} \in \tilde{V}}{\text{Min }} J(\tilde{v}).$$

– Show that a necessary condition allowing $\tilde{u}$ to be a solution of (**MP**) is written as:

Find the numerical sequence $(\tilde{u}_i)_{1, N+1}$ defining the approximation $\tilde{u}$ belonging to $\tilde{V}$, solution to the global linear system:

$$\sum_{j=1}^{N+1} A_{ij} \tilde{u}_j = b_i, \quad (\forall i = 1 \text{ to } N + 1), \tag{5.13}$$

where the following was noted:

$$A_{ij} \equiv ES \int_{Supp\ \varphi_i \cap Supp\ \varphi_j} \varphi_i' \varphi_j' \, dx, \quad b_i \equiv \int_{Supp\ \varphi_i} f \varphi_i \, dx. \tag{5.14}$$

8) The elements of the elementary matrix $a^{(i+1)}$, leading the contribution of the segment $[x_i, x_{i+1}]$, i.e. the $(i+1)$-th element of the mesh of the interval $[0, L]$, into the global matrix A of the linear system (5.13), having a generic element A_{ij} is now introduced. Thus, the following is written:

$$
a^{(i+1)} = \begin{bmatrix} a_{1,1}^{(i+1)} & a_{1,2}^{(i+1)} \\ a_{2,1}^{(i+1)} & a_{2,2}^{(i+1)} \end{bmatrix},
$$

$$
\equiv ES \begin{bmatrix} \displaystyle\int_{x_i}^{x_{i+1}} (\dot{\varphi}_i)^2 \, dx & \displaystyle\int_{x_i}^{x_{i+1}} \dot{\varphi}_i \dot{\varphi}_{i+1} \, dx \\ \displaystyle\int_{x_i}^{x_{i+1}} \dot{\varphi}_{i+1} \dot{\varphi}_i \, dx & \displaystyle\int_{x_i}^{x_{i+1}} (\dot{\varphi}_{i+1})^2 \, dx \end{bmatrix}. \tag{5.15}
$$

Likewise, introduce the elementary vector $b^{(i+1)}$ defined by:

$$
b^{(i+1)} = \begin{bmatrix} b_1^{(i+1)} \\ b_2^{(i+1)} \end{bmatrix} \equiv \begin{bmatrix} \displaystyle\int_{x_i}^{x_{i+1}} f \varphi_i \, dx \\ \displaystyle\int_{x_i}^{x_{i+1}} f \varphi_{i+1} \, dx \end{bmatrix}. \tag{5.16}
$$

Find the relationship between the coefficients A_{ij} and $a_{ij}^{(i+1)}$, and then, between b_i and $b^{(i)}$. Qualitatively explain the assembly technique inferred from it.

9) Exactly calculate the 4 coefficients of the elementary matrix $a^{(i+1)}$ then propose an approximation for the vector $b^{(i+1)}$ by using the trapezium rule.

10) Assemble the global matrix A having generic element A_{ij} and the second member b_i defined by the problem $(\widetilde{\mathbf{MP}})$.

11) Are the nodal equations obtained by the classic technique of approximation by finite elements $\mathbf{P_1}$ applied to the variational formulation $(\mathbf{VP})$ found again?

12) And what about the finite differences method applied to the continuous problem $(\mathbf{CP})$?

5.1.2 Case of an Elastic Support

13) The restraint condition when $x = 0$ is replaced by an elastic support having a given stiffness k (cf. Fig.5.2). When the spring of the support is at rest, the displacement u_1 at the end of the beam having abscissa $x_1 = 0$ is zero.

Knowing that the beam is subjected to the force density $\mathbf{f} = f_1 \mathbf{X}_1$, show that the minimization of the potential energy of the system "Beam-Spring" is written as:

$$(\textbf{PM}) \quad \begin{bmatrix} \text{Find } u \text{ belonging to } V', \text{ solution to:} \\ J(u) = \underset{v \in V'}{\text{Min }} J(v) \,, \\ \text{where:} \\ J(v) = \frac{ES}{2} \int_0^L v'^2(x) \, dx - \int_0^L f(x) v(x) \, dx + \frac{1}{2} k u_1^2 \,. \end{bmatrix} \qquad (5.17)$$

It may have been noticed that the displacement fields belonging to V' are no longer subjected to the homogenous Dirichlet condition when $x_1 = 0$.

14) The particular case of a mesh of the beam Ω, composed of only one mesh $[x_1 = 0, x_2 = L]$ is considered.

– Write the system of equations verified by the approximate displacements $\tilde{u}_1$ and $\tilde{u}_2$, (respectively at nodes x_1 and x_2).

– Solve the system of equations and find the approximate displacements $\tilde{u}_1$ and $\tilde{u}_2$ and propose an approximation of the displacement field $\tilde{u}$ at any point of the beam.

15) When the stiffness of the elastic support is infinite, show that the results of a restrained beam when $x_1 = 0$ is found again.

———————————

Solution

▶ **Displacement Variational Formulation – Principle of Virtual Work – Theoretical Part**

A.1) The choice of a virtual field defined by (5.1) is justified by the physical nature of the forces acting on the beam Ω.

Indeed, the beam is exclusively subject to a density **f** which prompts the search for a real displacement field **U**, and as a result, of virtual fields **U*** in the form of (5.1).

The expression of the strain tensor $\varepsilon(\mathbf{U}^*)$, associated with the virtual field **U***, can be easily inferred from the shape of the virtual fields (5.1) by using the definition (5.3):

$$\varepsilon(\mathbf{U}^*) = \varepsilon_{11}^* \mathbf{X}_1 \otimes \mathbf{X}_1 = \dot{u}^*(x) \mathbf{X}_1 \otimes \mathbf{X}_1 , \tag{5.18}$$

in which the tensor product $\mathbf{X}_1 \otimes \mathbf{X}_1$ indicates that the strain tensor field $\varepsilon(\mathbf{U}^*)$ has only one non-zero component, namely: ε_{11}^*. Therefore, the evaluation of the virtual work of internal forces can be performed:

$$T_{\text{Int}}^* = -\int_\Omega \sigma_{ij}\varepsilon_{ij}(\mathbf{U}^*)\,\mathrm{d}\Omega = -\int_\Omega \sigma_{11}\varepsilon_{11}^*\,\mathrm{d}\Omega \tag{5.19}$$

$$= -\int_0^L \left(\iint_{S(x)} \sigma_{11}\,\mathrm{d}S(x) \right) \dot{u}^*(x)\,\mathrm{d}x = -\int_0^L N(x)\dot{u}^*(x)\,\mathrm{d}x , \tag{5.20}$$

in which the expression of the normal forces N as defined by (5.4) is used.

In the same way, the virtual work of external forces is evaluated:

$$T_{\text{Ext}}^* = \int_\Omega \mathbf{f} \cdot \mathbf{U}^*\,\mathrm{d}\Omega = \int_\Omega f_1 u^*(x)\,\mathrm{d}\Omega \tag{5.21}$$

$$= \int_0^L \left(\iint_{S(x)} f_1\,\mathrm{d}S(x) \right) u^*(x)\,\mathrm{d}x = \int_0^L f(x)u^*(x)\,\mathrm{d}x . \tag{5.22}$$

A.2) The application of the principle of virtual work for static phenomena is written as:

$$T_{\text{Int}}^* + T_{\text{Ext}}^* = 0, \ \forall u^* \text{ which fulfils the conditions in (5.1).} \tag{5.23}$$

Fig. 5.2 Beam with Elastic Support

In other words, by using the expressions (5.20) and (5.21), the following is obtained:

$$-\int_0^L N(x)\dot{u}^*(x)\,dx + \int_0^L f(x)u^*(x)\,dx = 0, \quad \forall u^*/u^*(0) = 0. \qquad (5.24)$$

An integration by parts is then performed on the first integral of (5.24) and the homogenous Dirichlet condition as defined by (5.1) is used.

The variational problem (**EVP**) is obtained:

$$(\textbf{EVP}) \left[\begin{array}{l} \int_0^L \dot{N}(x)u^*(x)\,dx + N(L)u^*(L) + \int_0^L f(x)u^*(x)\,dx = 0, \\[2mm] \forall u^*/u^*(0) = 0. \end{array} \right. \qquad (5.25)$$

A.3) To obtain the formulation of the continuous problem (**CP**), the equation (5.25) of the variational problem (**EVP**) is again used and the normal force N is replaced by its expression in relation with the associated displacement field u, *via* the behaviour law, as pointed out in the statement of this question.

The following is then obtained:

$$(\textbf{EVP}) \left[\begin{array}{l} \int_0^L [ES\ddot{u}(x) + f(x)]\,u^*(x)\,dx + ES\dot{u}(L)u^*(L) = 0, \\[2mm] \forall u^*/u^*(0) = 0. \end{array} \right. \qquad (5.26)$$

From the variational formulation (5.26), which is in fact a formal formulation, a functional framework for the working out of solution u, will be proposed; i.e. a plausible existence space for u as well as for virtual displacement fields u^* which occur in (5.26).

To achieve this, considering that the density f is a given function belonging to $L^2(0,L)$, the application of the Cauchy-Schwartz inequality to the integral of equ. (5.26) ensures its existence, provided that $\ddot{u}$ and u^* belong to $L^2(0,L)$.

This is why the solution u and the virtual fields u^* are considered in Sobolev space $H^2(0,L)$.

It can further be remarked that the functions of $H^2(0,L)$ being C^1 on $[0,L]$ (cf. H. Brézis, [1]), the values of $\dot{u}$ and of u^* at the point $x = L$, occurring in formulation (5.26), are perfectly licit.

Moreover, the homogenous Dirichlet condition $u(0) = u^*(0) = 0$ leads to the consideration of the displacement field u, being the solution of (5.26) as well as the virtual displacement fields u^* in the space $H_*^2(0,L)$ as defined by:

$$H_*^2(0,L) \equiv \{v\colon [0,L] \to \mathbf{R},\ v \in H^2(0,L)\ \text{and}\ v(0) = 0\}. \qquad (5.27)$$

The variational formulation (**EVP**) is written as:

$$
\textbf{(EVP)} \quad
\begin{cases}
\text{Find } u \text{ belonging to } H^2_*(0,L) \text{ which is the solution to:} \\[2mm]
\displaystyle\int_0^L [ES\ddot{u}(x) + f(x)]\,u^*(x)\,\mathrm{d}x + ES\dot{u}(L)u^*(L) = 0\,, \\[2mm]
\forall u^* \in H^2_*(0,L)\,.
\end{cases}
\tag{5.28}
$$

The continuous problem (**CP**) is then solved in two steps:

Firstly, the equ. (5.28) is studied, the first particular case being that of the functions u^* belonging to $H^2_*(0,L)$ and which are moreover zero when $x = L$.

For such functions, equ. (5.28) is written as:

$$
\int_0^L [ES\ddot{u}(x) + f(x)]\,u^*(x)\,\mathrm{d}x = 0\,, \quad \forall u^* \in H^2_*(0,L)\,/\,u^*(L) = 0\,.
\tag{5.29}
$$

The process is then performed using density while noting that (5.29) is also verified for any function u^* belonging to $\mathscr{D}(0,L)$ which comprises the function spaces C^∞ having compact support as

$$
\mathscr{D}(0,L) \subset H^2_*(0,L)\,.
$$

The fact that $\mathscr{D}(0,L)$ is dense in $L^2(0,L)$ is then used:

$$
\forall g \in L^2(0,L),\ \exists g_n \in \mathscr{D}(0,L) \quad \text{such that: } \lim_{n \to +\infty} \|g_n - g\|_{L^2} = 0\,,
\tag{5.30}
$$

the limit being considered as for the L^2-norm.

Thus, when g is fixed, (while remaining an undefined value), in $L^2(0,L)$, the sequence g_n of $\mathscr{D}(0,L)$ defined by (5.30), verifies the variational equ. (5.29) like any function of $\mathscr{D}(0,L)$.

In fact, it would be convenient for (5.29) to be verified by the function g belonging to $L^2(0,L)$ so as to choose among all the g functions of $L^2(0,L)$ the precise function that is equal to the specific function G defined by:

$$
G \equiv ES\ddot{u}(x) + f(x)\,.
$$

This easily yields the differential equation of the continuous problem (**CP**).

To achieve this, the left member of equ. (5.29) is checked and if u^* is a function g of $L^2(0,L)$.

$$
\left| \int_0^L G(x)g(x)\,\mathrm{d}x \right| \leq \int_0^L |G(x)| \cdot |g(x) - g_n(x)|\,\mathrm{d}x\,,
$$

$$
\leq \left[\int_0^L G^2(x)\,\mathrm{d}x \right]^{1/2} \left[\int_0^L |g(x) - g_n(x)|^2(x)\,\mathrm{d}x \right]^{1/2}\,.
\tag{5.31}
$$

The limit process in the inequality (5.31) is then done and this leads to the variational equality (5.29), for any function g belonging to $L^2(0,L)$.

As mentioned earlier, the differential equation of the continuous problem (**CP**) is then obtained immediate; it only requires the choice of a function among all the functions g belonging to $L^2(0,L)$ that satisfies (5.29) and that strictly equals G, $(g \equiv G)$.

The variational equality (5.28) that, as shown earlier, holds a zero integral in so far as its integrand $[ES\ddot{u}(x) + f(x)]$ is necessarily zero, is processed again.

It then becomes:

$$\dot{u}(L)u^*(L) = 0, \quad \forall u^* \in H_*^2(0,L) .\tag{5.32}$$

It can be easily concluded that $\dot{u}(L) = 0$.

▶ **Summary**

The solution u of the variational equation (5.28) has been worked out in the space $H_*^2(0,L)$, as defined in (5.27).

In so far as this particular solution, as shown earlier, satisfies the differential equation of the problem (**CP**), on the one hand and the Neumann condition $\dot{u}(L) = 0$ on the other hand and even the homogenous Dirichlet condition with $x = 0$, as a property of the space $H_*^2(0,L)$, it proves that the function u is the exact solution of the continuous problem (**CP**), as defined in (5.6).

A.4) Considering the way in which the different formulations occurring in the given problem based on mechanical considerations were constructed, the variational formulation (**VP**) that may be obtained through the traditional method as proposed in the previous chapters is used again as follows.

Indeed, it only suffices to revert to formulation (5.28) keeping in mind that the functional framework used for this formulation has been defined in the previous question, namely: $H_*^2(0,L)$.

Moreover, should u be a solution to the continuous problem (**CP**) and u satisfies the Neumann condition $\dot{u}(L) = 0$, equ. (5.28) is re-written as:

Find u belonging to $H_*^2(0,L)$ which is the solution of:

$$\int_0^L [ES\ddot{u}(x) + f(x)]\, u^*(x)\, dx = 0, \quad \forall u^* \in H_*^2(0,L) .\tag{5.33}$$

An integration by parts in variational equation (5.33), coupled with the boundary conditions satisfied by u^* when $x = 0$ and by when $x = L$, enables establishment of the variational formulation (**VP**) as defined in (5.8).

Concerning the functional framework of this formulation, subsequent to the integration by parts just mentioned, a degree of derivation is "lost" in the course of this transformation.

This is why the problem (**VP**) is set down in $H^1_*(0,L)$, so as to ensure convergence of the integrals found in its formulation.

A.5) In order to establish the equivalence between the variational formulation (**VP**) and the minimisation problem (**MP**) defined in (5.9), it is only necessary to note that the variational formulation (**VP**) is written as:

$$a(u,v) = L(v), \quad \forall v \in V, \tag{5.34}$$

in which the bilinear form $a(.,.)$ and the linear form $L(.)$ satisfy the properties ensuring the equivalence between (**VP**) and (**MP**)

It principally concerns the symmetry and positivity of form $a(.,.)$, (cf. D. Euvrard, [4] or P.A. Raviard, [7]).

Indeed, it suffices to state the formula below in the framework of the variational formulation (5.8):

$$V \equiv H^1_*(0,L), \tag{5.35}$$

$$a(u,v) \equiv ES \int_0^L \dot{u}(x)\dot{v}(x)\,dx, \tag{5.36}$$

$$L(v) \equiv \int_0^L f(x)v(x)\,dx. \tag{5.37}$$

▶ *Remark*

The minimisation problem (**MP**) which is defined by (5.9) gives an interpretation from a mechanical point of view.

Indeed, the function J to be minimised on the set of the displacement fields $\mathbf{v} = v(x)\mathbf{X}_1$, is none other than the potential energy E_p of the beam Ω, as defined by:

$$E_p(\mathbf{v}) \equiv E_{\mathrm{Def}}(\mathbf{v}) - W_{\mathrm{Ext}}(\mathbf{v}), \tag{5.38}$$

in which $E_{\mathrm{Def}}(\mathbf{v})$ is the modulus of resilience corresponding to:

$$E_{\mathrm{Def}}(\mathbf{v}) \equiv \frac{1}{2}\int_\Omega \sigma_{ij}\varepsilon_{ij}\,d\Omega = \frac{1}{2}\int_\Omega \sigma_{11}\varepsilon_{11}\,d\Omega = \frac{E}{2}\int_\Omega \varepsilon_{11}^2\,d\Omega, \tag{5.39}$$

$$= \frac{E}{2}\int_0^L \left(\iint_{S(x)} dS(x)\right) \dot{v}^2\,dx = \frac{ES}{2}\int_0^L \dot{v}^2\,dx. \tag{5.40}$$

Furthermore, $W_{\text{Ext}}(\mathbf{v})$ represents the known work of forces in the unknown displacement $\mathbf{v}$, namely:

$$W_{\text{Ext}}(\mathbf{v}) \equiv \int_{\Omega} \mathbf{f} \cdot \mathbf{v} \, d\Omega = \int_0^L \left(\iint_{S(x)} f_1 \, dS(x) \right) v(x) \, dx ,$$

$$= \int_0^L f(x) v(x) \, dx . \tag{5.41}$$

Then it suffices to replace the respective expressions of (5.40) and (5.41) in the definition of the potential energy $E_p(\mathbf{v})$, which is defined in (5.38), so as to precisely find the definition of the function J of the problem (**MP**), (cf. (5.9)).

▶ **Numerical Part**

A.6) From this question onwards, the approximation of the minimisation problem (**MP**) and, in a similar manner, that of the variational formulation (**VP**) would be performed using finite elements $\mathbf{P_1}$.

In this perspective, any function $\tilde{v}$ belonging to the approximation space $\tilde{V}$ is factorized on the canonical basis form made up of functions $(\varphi_i)_{i=1,N+1}$, satisfying:

$$\varphi_i(x_j) = \delta_{ij} . \tag{5.42}$$

The decomposition of any element $\tilde{v}$ of $\tilde{V}$ is then written as:

$$\tilde{v} = \sum_{j=1}^{N+1} \tilde{v}_j \varphi_j . \tag{5.43}$$

Considering the property (5.42), if $\tilde{v}$ defined by (5.43) is estimated at the abscissa x_i, the following is obtained:

$$\tilde{v}(x_i) = \sum_{j=1}^{N+1} \tilde{v}_j \varphi_j(x_i) = \sum_{j=1}^{N+1} \tilde{v}_j \delta_{ij} = \tilde{v}_i , \tag{5.44}$$

where the properties of the Krönecker symbol δ_{ij} would have been used:

$$\delta_{ij} = \begin{vmatrix} 1, \text{ if } i = j, \\ 0, \text{ if } i \neq j . \end{vmatrix} \tag{5.45}$$

Thus, $\tilde{v}_i$ is interpreted exactly as the value of an approximate displacement field $\tilde{v}$ at the point of discretisation x_i.

For this reason, and this constitutes the major consequence of the choice of the basis functions $(\varphi_i)_{i=1,N+1}$, which satisfy the conditions (5.42), the unknown coefficients $(\tilde{u}_1, \tilde{u}_2, \ldots, \tilde{u}_{N+1})$ in the linear combination defining the approximate solution $\tilde{u}$, correspond to the approximation of $(N+1)$ approximate displacements $(\tilde{u}(x_1), \tilde{u}(x_2), \ldots, \tilde{u}(x_{N+1}))$ of the displacement field $\tilde{u}$ at the nodes of the mesh $(x_1, x_2, \ldots, x_{N+1})$.

A.7) The approximation of the minimisation problem of problem (**MP**) can now be worked out. To this end, any function v in the functional space V is replaced by its approximation $\tilde{v}$ belonging to the space $\tilde{V}$ which is defined by (5.11).

The approximate minimisation problem $(\widetilde{\textbf{MP}})$ is therefore written as:

$$(\widetilde{\textbf{MP}}) \quad \begin{bmatrix} \text{Find } \tilde{u} \text{ belonging to } \tilde{V} \text{ which is the solution to:} \\[2mm] J(\tilde{u}) = \underset{\tilde{v} \in \tilde{V}}{\text{Min}} \ J(\tilde{v}) , \\[3mm] \text{with}: \quad J(\tilde{v}) = \frac{ES}{2} \int_0^L \left[\sum_{j=1}^{N+1} \tilde{v}_j \dot{\varphi}_j \right]^2 dx - \int_0^L f \left[\sum_{j=1}^{N+1} \tilde{v}_j \varphi_j \right] dx . \end{bmatrix} \tag{5.46}$$

In this form, the functional J can be considered as a function of $(N+1)$ variables $(\tilde{v}_1, \tilde{v}_2, \ldots, \tilde{v}_{N+1})$.

Therefore, a necessary condition of minimisation of J is written as:

$$\frac{\partial J}{\partial \tilde{v}_j}(\tilde{u}_1, \tilde{u}_2, \ldots, \tilde{u}_{N+1}) = 0, \quad \forall j = 1, \ldots, N+1 . \tag{5.47}$$

The expression (5.46) is then used again and partial derivation is applied in relation with each $\tilde{v}_j$, $(j = 1 \text{ to } N+1)$.

$$(\widetilde{\textbf{MP}}) \quad \begin{bmatrix} \text{For all } i \text{ belonging to } \{1, \ldots, N+1\}, \text{ it follows that:} \\[2mm] \left(\frac{\partial J}{\partial \tilde{v}_j}(\tilde{u}_1, \tilde{u}_2, \ldots, \tilde{u}_{N+1}) = 0 \right) \\[3mm] \Leftrightarrow \left(ES \sum_{j=1}^{N+1} \left[\int_0^L \dot{\varphi}_i \dot{\varphi}_j \, dx \right] \tilde{u}_j = \int_0^L f(x) \varphi_i \, dx \right) . \end{bmatrix} \tag{5.48}$$

Hence, the linear system (5.13) is obtained.

A.8) Using simple identification, the relationships between coefficients $a_{i,j}^{(i+1)}$ and $A_{i,j}$ are obtained, provided that it is shown that only the coefficients $A_{i,i-1}$, $A_{i,i}$ and $A_{i,i+1}$ are non-zero, a priori, in the global matrix A.

These relationships are then expressed as:

$$A_{i,i-1} = ES \int_{Supp\ \varphi_{i-1}\ \cap\ Supp\ \varphi_i} \dot{\varphi}_{i-1} \dot{\varphi}_i \, dx = ES \int_{x_{i-1}}^{x_i} \dot{\varphi}_{i-1} \dot{\varphi}_i \, dx \equiv a_{2,1}^{(i)}, \tag{5.49}$$

$$A_{i,i} = ES \int_{Supp\ \varphi_i} \dot{\varphi}_i^2 \, dx = ES \left[\int_{x_{i-1}}^{x_i} \dot{\varphi}_i^2 \, dx + \int_{x_i}^{x_{i+1}} \dot{\varphi}_i^2 \, dx \right] \equiv a_{2,2}^{(i)} + a_{1,1}^{(i+1)}, \tag{5.50}$$

$$A_{i,i+1} = ES \int_{Supp\ \varphi_i\ \cap\ Supp\ \varphi_{i+1}} \dot{\varphi}_i \dot{\varphi}_{i+1} \, dx = ES \int_{x_i}^{x_{i+1}} \dot{\varphi}_i \dot{\varphi}_{i+1} \, dx \equiv a_{1,2}^{(i+1)} . \tag{5.51}$$

Having established the three relationships (5.49), (5.50) and (5.51) between the coefficients of the global matrix A and those of the local matrix $a^{(i+1)}$, showing the contribution of each element $[x_i, x_{i+1}]$ in matrix A, it appears that each A_{ij} coefficient is "made up" of either one or two geometrical finite elements contributions.

Moreover, express the ith and $(i+1)^{\text{th}}$ lines of linear system (5.13) using the coefficients of the local matrix $a^{(i)}$.

The result is:

$$a_{2,1}^{(i)} \tilde{u}_{i-1} + \left[a_{2,2}^{(i)} + a_{1,1}^{(i+1)} \right] \tilde{u}_i + a_{1,2}^{(i+1)} \tilde{u}_{i+1} = b_i , \tag{5.52}$$

$$a_{2,1}^{(i+1)} \tilde{u}_i + \left[a_{2,2}^{(i+1)} + a_{1,1}^{(i+2)} \right] \tilde{u}_{i+1} + a_{1,2}^{(i+2)} \tilde{u}_{i+2} = b_{i+1} , \tag{5.53}$$

where b_i and b_{i+1} are noted as second member of system (5.13).

In addition, there is:

$$b_i \equiv \int_0^L f(x)\varphi_i \, dx = \int_{x_{i-1}}^{x_{i+1}} f(x)\varphi_i \, dx \tag{5.54}$$

$$= \int_{x_{i-1}}^{x_i} f(x)\varphi_i \, dx + \int_{x_i}^{x_{i+1}} f(x)\varphi_i \, dx \equiv b_2^{(i)} + b_1^{(i+1)} , \tag{5.55}$$

where definition (5.16) of the elementary vector $b^{(i+1)}$ has been used.

Thus $b^{(i+1)}$ represents the contribution of mesh $[x_i, x_{i+1}]$ in the constitution of the second member b_i just as the local matrix $a^{(i+1)}$ does for the global matrix A.

The contribution of mesh $[x_i, x_{i+1}]$ has thus been identified for the constitution of the global matrix A and for the second member b.

Moreover, concerning the elements of the local matrix $a^{(i+1)}$, these weight the approximate displacements $\tilde{u}_i$ and $\tilde{u}_{i+1}$ as follows:

$$
\begin{array}{cc}
\tilde{u}_i & \tilde{u}_{i+1} \\
\downarrow & \downarrow
\end{array}
$$

$$
\begin{array}{c}
\tilde{u}_i \rightarrow \\
\tilde{u}_{i+1} \rightarrow
\end{array}
\begin{bmatrix}
a_{1,1}^{(i+1)} & a_{1,2}^{(i+1)} \\
a_{2,1}^{(i+1)} & a_{2,2}^{(i+1)}
\end{bmatrix} ,
\begin{bmatrix}
b_1^{(i+1)} \\
b_2^{(i+1)}
\end{bmatrix} . \tag{5.56}
$$

The technique of assembly then consists in passing each geometrical element $[x_i, x_{i+1}]$ one by one, while transcribing again the contribution of each mesh in the global matrix A and in the second member b of the linear system (5.13).

This assembly methodology is licit since it relies on the linear structure of the global matrix A and of the second member b from the point of view of contributions of elementary meshes $[x_i, x_{i+1}]$.

This is none other than the qualitative representation of equs. (5.52) to (5.55) that brings out the different above-mentioned contributions by "linear combinations".

Start assembling with the first element $[x_0, x_1]$. If node x_0 is restrained $(u(0) = 0)$, the degree of freedom corresponding to $\tilde{u}_0$ is zero.

In other words, in this case, the elementary matrix $a^{(1)}$ is degenerate.

In fact, in this particular case, the following is obtained:

$$\left[a^{(1)} \right] (\tilde{u}_1) \equiv a_{2,2}^{(1)} \, \tilde{u}_1 \,. \tag{5.57}$$

Similarly, for the second member b, the contribution of mesh $[x_0, x_1]$ is that of node x_1. This contribution is exactly equivalent to $b_2^{(1)}$ as defined in (5.16).

Using this first result, the coefficient $a_{2,2}^{(1)}$ is inserted in the global matrix A by positioning it at the corresponding place, i. e. in the first line of the first column.

The same process is used to integrate the contribution of coefficient $b_2^{(1)}$ in the second member:

$$\begin{bmatrix} a_{2,2}^{(1)} & 0 & \dots & \dots & 0 \\ 0 & 0 & \dots & \dots & 0 \\ \dots & \dots & \dots & \dots & \dots \\ \dots & \dots & \dots & \dots & \dots \\ 0 & 0 & \dots & \dots & 0 \end{bmatrix}, \quad \begin{bmatrix} b_2^{(1)} \\ \dots \\ \dots \\ \dots \\ \dots \end{bmatrix}. \tag{5.58}$$

The second element $[x_1, x_2]$ is now examined. Its contribution in the global matrix is determined by the elementary matrix $\mathbf{a}^{(2)}$ affected by the corresponding approximate displacements $\tilde{u}_1$ and $\tilde{u}_2$:

$$\begin{array}{cc} \tilde{u}_1 & \tilde{u}_2 \\ \downarrow & \downarrow \end{array}$$

$$\begin{array}{c} \tilde{u}_1 \rightarrow \\ \tilde{u}_2 \rightarrow \end{array} \begin{bmatrix} a_{1,1}^{(2)} & a_{1,2}^{(2)} \\ a_{2,1}^{(2)} & a_{2,2}^{(2)} \end{bmatrix}, \quad \begin{bmatrix} b_1^{(2)} \\ b_2^{(2)} \end{bmatrix}. \tag{5.59}$$

The contribution of the second mesh $[x_1, x_2]$ is then inserted in the global matrix A by affecting the coefficients of elementary matrix $\mathbf{a}^{(2)}$ (5.59) at the relevant places in the matrix A.

The same process is applied to the second member b and the result is:

$$
\begin{bmatrix}
(a_{2,2}^{(1)}+\mathbf{a}_{1,1}^{(2)}) & \mathbf{a}_{1,2}^{(2)} & \cdots & \cdots & 0 \\
\mathbf{a}_{2,1}^{(2)} & \mathbf{a}_{2,2}^{(2)} & \cdots & \cdots & 0 \\
\cdots & \cdots & \cdots & \cdots & \cdots \\
\cdots & \cdots & \cdots & \cdots & \cdots \\
0 & 0 & \cdots & \cdots & 0
\end{bmatrix}
,\qquad
\begin{bmatrix}
b_2^{(1)}+\mathbf{b}_1^{(2)} \\
\mathbf{b}_2^{(2)} \\
\cdots \\
\cdots \\
\cdots
\end{bmatrix}
. \tag{5.60}
$$

The global matrix A and the second member b are thus filled hop-by-hop by passing each element of the mesh, one by one, in order to attach the contribution of each mesh $[x_i, x_{i+1}]$ by using the corresponding elementary matrix $a^{(i+1)}$ and the local second member $b^{(i+1)}$.

The final result corresponds to the following matrix A:

$$
A =
\begin{bmatrix}
(a_{2,2}^{(1)}+a_{1,1}^{(2)}) & a_{1,2}^{(2)} & \cdots & \cdots & & 0 \\
a_{2,1}^{(2)} & (a_{2,2}^{(2)}+a_{1,1}^{(3)}) & a_{1,2}^{(3)} & \cdots & & 0 \\
0 & a_{2,1}^{(3)} & (a_{2,2}^{(3)}+a_{1,1}^{(4)}) & \cdots & & \cdots \\
\cdots & \cdots & \cdots & \cdots & & \cdots \\
0 & 0 & (a_{2,2}^{(N-1)}+a_{1,1}^{(N)}) & a_{1,2}^{(N)} & & \cdots \\
\cdots & \cdots & a_{2,1}^{(N)} & (a_{2,2}^{(N)}+a_{1,1}^{(N+1)}) & a_{1,2}^{(N+1)} \\
0 & 0 & \cdots & a_{2,1}^{(N+1)} & (a_{1,1}^{(N)}+a_{2,2}^{(N+1)})
\end{bmatrix}
\tag{5.61}
$$

and to second member b:

$$
b =
\begin{bmatrix}
b_2^{(1)}+b_1^{(2)} \\
b_2^{(2)}+b_1^{(3)} \\
\cdots \\
\cdots \\
\cdots \\
b_2^{(N)}+b_1^{(N+1)} \\
b_2^{(N+1)}
\end{bmatrix}
. \tag{5.62}
$$

A.9) Calculation of the matrix coefficients $a^{(i+1)}$ is performed without approximation, provided that the integrands are solely composed of derivatives of affine functions per mesh $[x_i, x_{i+1}]$. In other words, the integrands bear on constant functions by mesh.

It then becomes:

$$a_{1,1}^{(i+1)} = ES \int_{x_i}^{x_{i+1}} (\dot{\phi}_i)^2 \, dx = \frac{ES}{h} \,, \tag{5.63}$$

$$a_{1,2}^{(i+1)} = a_{2,1}^{(i+1)} = ES \int_{x_i}^{x_{i+1}} \dot{\phi}_i \dot{\phi}_{i+1} \, dx = -\frac{ES}{h} \,, \tag{5.64}$$

$$a_{2,2}^{(i+1)} = ES \int_{x_i}^{x_{i+1}} (\dot{\phi}_{i+1})^2 \, dx = \frac{ES}{h} \,. \tag{5.65}$$

The elementary vector $b^{(i+1)}$ is now approximated using the trapezium quadrature formula.

$$b_1^{(i+1)} = \int_{x_i}^{x_{i+1}} f\varphi_i \, dx \simeq \frac{h}{2} \left[f(x_i)\varphi_i(x_i) + f(x_{i+1})\varphi_i(x_{i+1}) \right]$$

$$\simeq \frac{h}{2} \left[(f_i \times 1) + (f_{i+1} \times 0) \right] = \frac{h}{2} f_i \,. \tag{5.66}$$

$$b_2^{(i+1)} = \int_{x_i}^{x_{i+1}} f\varphi_{i+1} \, dx \simeq \frac{h}{2} \left[f(x_i)\varphi_{i+1}(x_i) + f(x_{i+1})\varphi_{i+1}(x_{i+1}) \right]$$

$$\simeq \frac{h}{2} \left[(f_i \times 0) + (f_{i+1} \times 1) \right] = \frac{h}{2} f_{i+1} \,. \tag{5.67}$$

A.10) After replacement of coefficients (5.63)–(5.65) of the elementary matrix and of approximations (5.66)–(5.67) of the constituents of the elementary vector in the structure of matrix A and in that of second member b previously obtained in (5.61) and (5.62), the following final result is obtained:

$$A = ES \begin{bmatrix} \frac{2}{h} & -\frac{1}{h} & 0 & \cdots & \cdots & \cdots & \cdots & 0 \\ -\frac{1}{h} & \frac{2}{h} & -\frac{1}{h} & 0 & \cdots & \cdots & \cdots & 0 \\ 0 & -\frac{1}{h} & \frac{2}{h} & -\frac{1}{h} & 0 & \cdots & \cdots & 0 \\ \cdots & \cdots & \cdots & \cdots & \cdots & \cdots & \cdots & \cdots \\ \cdots & \cdots & \cdots & \cdots & \cdots & \cdots & \cdots & \cdots \\ 0 & \cdots & \cdots & 0 & -\frac{1}{h} & \frac{2}{h} & -\frac{1}{h} & 0 \\ 0 & \cdots & \cdots & \cdots & 0 & -\frac{1}{h} & \frac{2}{h} & -\frac{1}{h} \\ 0 & \cdots & \cdots & \cdots & \cdots & 0 & -\frac{1}{h} & \frac{1}{h} \end{bmatrix} \,, \quad b \simeq \begin{bmatrix} hf_1 \\ hf_2 \\ hf_3 \\ \cdots \\ \cdots \\ hf_{N-1} \\ hf_N \\ \frac{h}{2} f_{N+1} \end{bmatrix} \,. \tag{5.68}$$

The "border effects" of the assembly techniques are noticed at the level of matrix A and of the second member b.

Namely, the last coefficient of diagonal $A_{N+1,N+1}$ is equal to $\dfrac{ES}{h}$ while the other diagonal coefficients are equivalent to $\dfrac{2ES}{h}$.

This is due to the fact that coefficient $A_{N+1,N+1}$ solely beneficiates from coefficient $a_{2,2}^{(N+1)}$ of the elementary matrix $a^{(N+1)}$ while the other diagonal coefficients $A_{i,i}$ are expressed in the form of $a_{2,2}^{(i)} + a_{1,1}^{(i+1)}$.

A.11) In order to compare the results obtained from the previous question (5.68) with those that could possibly come from the approximation of the variational formulation (**VP**) by finite elements $\mathbf{P_1}$, it must be noted that the approximate minimization (5.48) problem composed of linear system (5.13) is exactly identical to the one obtained from approximation of the (**VP**) formulation.

To ensure matters, retrieve expression (5.8) of the variational formulation (**VP**) and execute the classical approximation substitutions:

$$u(x) \rightarrow \tilde{u}(x) = \sum_{j=1}^{N+1} \tilde{u}_j \varphi_j , \qquad (5.69)$$

$$v(x) \rightarrow \tilde{v}(x) = \varphi_i . \qquad (5.70)$$

Thus, having the same system of equations defined by matrix A and the second member b (5.68), the nodal equation characteristic of a node strictly interior at x_i, $(i = 1, N)$ match the N first lines of the linear system (5.13) whose generic expression is:

$$\frac{ES}{h}[-\tilde{u}_{i-1} + 2\tilde{u}_i - \tilde{u}_{i+1}] = hf_i . \qquad (5.71)$$

The nodal equation corresponding to the basis function φ_{N+1} compares to the last equation of the linear system (5.13), i. e.:

$$\frac{ES}{h}[-\tilde{u}_N + \tilde{u}_{N+1}] = \frac{h}{2}f_{N+1} . \qquad (5.72)$$

A.12) The comparison is now examined with the finite differences method. It is immediately obvious that the nodal equation (5.71) is exactly identical to the one that would be obtained by discretisation of the differential equation of the continuous problem (**CP**) by finite differences.

As for the nodal equation (5.72) and its comparison with the discretisation of the Neumann condition $\dot{u}(L) = 0$, perform Taylor's expansion of solution u of the continuous problem (**CP**) – by assuming that it shows "sufficient" regularity around abscissa $x = L$.

$$u(x_N) = u(x_{N+1}) - h\dot{u}(x_{N+1}) + \frac{h^2}{2}\ddot{u}(x_{N+1}) + O(h^3), \qquad (5.73)$$

$$u(x_N) = u(x_{N+1}) - \frac{h^2}{2ES}f(x_{N+1}) + O(h^3), \qquad (5.74)$$

where the second derivative $\ddot{u}$ has been replaced by its expression obtained from the differential equation of the (**CP**) problem.

From equ. (5.74), consider approximations $\tilde{u}_i$ by omitting $O(h^3)$ and finally the nodal equation (5.72) is obtained.

▶ **Case for an Elastic Support**

A.13) When replacing the restraint at $x = 0$ by an elastic support having stiffness k, an additional kinematic degree of freedom needs to be introduced at this end of the beam. This justifies why u_1 is used to refer to the displacement at this end of the beam in relation to the fixed frame.

Moreover, the deformation work of the "Beam/Spring" system is expressed as:

$$E_{\text{Def}}(Beam/Spring) = E_{\text{Def}}(Beam) + E_{Def}(Spring). \qquad (5.75)$$

But each of these deformation works is expressed as:

$$E_{\text{Def}}(Beam) = \frac{ES}{2}\int_0^L \dot{u}^2(x)\,dx, \quad E_{\text{Def}}(Spring) = \frac{1}{2}ku_1^2, \qquad (5.76)$$

where the expression of the deformation work of a push-pull beam, established in the previous part (cf. (5.39)–(5.40)), is used directly.

The formulation of minimisation problem (5.71) is immediate as soon as it is shown that the Dirichlet condition, homogenous when $x_1 = 0$, is obsolete in defining the search space V'.

A.14) In the particular case of a beam meshed with a single element $[x_1 = 0, x_2 = L]$, the approximate solution is found on the whole beam in all displacement fields expressed as:

$$\tilde{v}(x) = \tilde{v}_1\varphi_1(x) + \tilde{v}_2\varphi_2(x). \qquad (5.77)$$

a) Equation system verified by approximate displacements $\tilde{u}_1$ and $\tilde{u}_2$.

The potential energy of minimisation problem, (5.17) measured on the displacement field defined by (5.77), is expressed as:

$$J(\tilde{v}) = \frac{ES}{2}\int_0^L (\tilde{v}_1\dot{\varphi}_1 + \tilde{v}_2\dot{\varphi}_2)^2\,dx - \int_0^L f(\tilde{v}_1\varphi_1 + \tilde{v}_2\varphi_2)\,dx + \frac{1}{2}k\tilde{v}_1^2. \qquad (5.78)$$

Then write both necessary minimisation conditions of the J functional as:

$$\frac{\partial J}{\partial \tilde{v}_1}(\tilde{u}_1, \tilde{u}_2) = \frac{\partial J}{\partial \tilde{v}_2}(\tilde{u}_1, \tilde{u}_2) = 0 .$$ (5.79)

Considering (5.78), these two conditions are expressed as:

$$\frac{\partial J}{\partial \tilde{v}_1}(\tilde{u}_1, \tilde{u}_2) = ES \int_0^L [\tilde{u}_1 \dot{\varphi}_1 + \tilde{u}_2 \dot{\varphi}_2] \dot{\varphi}_1 \, dx - \int_0^L f \varphi_1 \, dx + k \tilde{u}_1 = 0 ,$$ (5.80)

$$\frac{\partial J}{\partial \tilde{v}_2}(\tilde{u}_1, \tilde{u}_2) = ES \int_0^L [\tilde{u}_1 \dot{\varphi}_1 + \tilde{u}_2 \dot{\varphi}_2] \dot{\varphi}_2 \, dx - \int_0^L f \varphi_2 \, dx = 0 .$$ (5.81)

The linear system with unknowns $\tilde{u}_1$ and $\tilde{u}_2$ is then expressed as:

$$\begin{bmatrix} k + ES \int_0^L \dot{\varphi}_1^2 & ES \int_0^L \dot{\varphi}_1 \dot{\varphi}_2 \\ ES \int_0^L \dot{\varphi}_1 \dot{\varphi}_2 & ES \int_0^L \dot{\varphi}_2^2 \end{bmatrix} \begin{bmatrix} \tilde{u}_1 \\ \tilde{u}_2 \end{bmatrix} = \begin{bmatrix} b_1^{(2)} \\ b_2^{(2)} \end{bmatrix} .$$ (5.82)

b) *Determination of approximate displacements $\tilde{u}_1$ and $\tilde{u}_2$ and approximation of the global displacement field along the beam.*

Solving the linear system (5.82) leads to solution:

$$\tilde{u}_1 = \frac{L}{2k}(f_1 + f_2), \quad \tilde{u}_2 = \frac{L^2}{2ES}f_2 + \frac{L}{2k}(f_1 + f_2) .$$ (5.83)

Thus, the approximate displacement $\tilde{u}$ on the whole beam is obtained by injecting the values of (5.83) in equation (5.77):

$$\tilde{u}(x) = \left[\frac{L}{2k}(f_1 + f_2) \right] \varphi_1(x) + \left[\frac{L^2}{2ES}f_2 + \frac{L}{2k}(f_1 + f_2) \right] \varphi_2(x) .$$ (5.84)

It is to be noted that the approximate displacement fields $\tilde{u}$ defined by (5.84) provide an approximation of the displacement field u on any point defined by $[0, L]$, through the usual finite element method.

It is one of the main differentiation points, in relation to the finite differences method, producing a sequence of approximations at points fixed on a discretisation grid, namely on the nodes of a predefined mesh of the beam.

A.15) The boundary case of a restraint when $x_1 = 0$, tackled from the elastic support angle having a coefficient of stiffness k tending towards infinity, is obtained by a run at the boundary in the expressions of $\tilde{u}_1$ and $\tilde{u}_2$ defined by (5.83). Thus, the asymptotic approximations $\tilde{u}_1^\infty$ and $\tilde{u}_2^\infty$ are expressed as:

$$\tilde{u}_1^\infty = 0, \quad \tilde{u}_2^\infty = \frac{L^2}{2ES}f_2 .$$ (5.85)

Compare these two values to those that could possibly be obtained directly from the problem of the fully fixed beam studied in the previous section.

To do this, just revert to the global matrix as well as to the second member (5.68) and adapt them to a case of a single mesh $[x_1, x_2]$, restrained at $x_1 = 0$.

A degenerated system is thus obtained from a single equation containing a single unknown $\tilde{u}_2$, such as:

$$\left(\frac{ES}{L}\right) \times \tilde{u}_2 = \frac{L}{2} f_2 . \tag{5.86}$$

The same expression obtained for $\tilde{u}_2$ is obtained for $\tilde{u}_2^\infty$ and is defined by (5.85):

$$\tilde{u}_2 = \frac{L^2}{2ES} f_2 . \tag{5.87}$$

Finally, it is noted that, in cases of restraint, the boundary conditions are obtained when $x_1 = 0$: $\tilde{u}_1 = \tilde{u}_1^\infty = 0$.

5.2 Beam Subject to Simple Bending

5.2.1 Beam Fitted With a Restraint and Having a Freely Movable Bearing

Statement

Consider a homogenous beam Ω of length L, cross-section S and constant density ρ, where the isotropic elasticity with small disturbances constitutes the mechanical behaviour. Moreover, it is noted that the constituent material of the beam has a Young's modulus of E and a Poisson's ratio of v.

The beam's axis is designated by $(O; \mathbf{X}_1)$ and the current abscissa is x. The beam's cross-section is parameterised by the geometric variables (y, z).

Moreover, the beam's geometric symmetry as well as that of the forces to which it is subjected enables considering the plan $(O; \mathbf{X}_1, \mathbf{X}_2)$ as the symmetrical plan.

The beam is restrained at its end where $x = 0$ and rests on a freely movable bearing restricting any vertical movement when $x = L$.

Moreover, a volume density of the transverse forces $\mathbf{f} = -f_2 \mathbf{X}_2$ is applied along the beam (see Fig. 5.3).

Density f_2 is given and has all properties of functional regularity so that the integration calculations of the first two questions of the theoretical part may be performed.

▶ **Displacement Variational Formulation – Virtual Work Principle**
  – Theoretical Part

1) The kinematic framework retained to describe the real displacement field is the Timoshenko theory. Thus, the virtual displacement fields $\mathbf{V}^*$ to be considered are written as:

$$\mathbf{V}^* = -y\Omega^*(x)\mathbf{X}_1 + v^*(x)\mathbf{X}_2 , \tag{5.88}$$

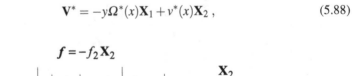

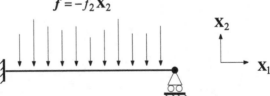

Fig. 5.3 Fixed Beam Resting on a Freely Movable Bearing

satisfying boundary conditions:

$$v^*(0) = \Omega^*(0) = 0 \quad \text{and} \quad v^*(L) = 0. \tag{5.89}$$

In addition, the boundary conditions (5.89) show, on one part, the presence of a restraint when $x = 0$, and on the other part, that of a freely movable bearing restricting any vertical displacement when $x = L$ for any virtual displacement field $\mathbf{V}^*$.

Thus $\mathfrak{V}^*$ denotes the set of the virtual displacement fields satisfying properties (5.88)–(5.89).

Determine the virtual work of internal forces T_{Int}^* as well as the virtual work of external forces T_{Ext}^* defined by:

$$T_{\text{Int}}^* = -\int_\Omega \sigma_{ij}\varepsilon_{ij}(\mathbf{V}^*)\,d\Omega, \quad T_{\text{Ext}}^* = \int_\Omega \mathbf{f}\cdot\mathbf{V}^*\,d\Omega, \tag{5.90}$$

where σ denotes the stress tensor and $\varepsilon(\mathbf{V}^*)$ the linear strain tensor associated with the virtual displacement fields $\mathbf{V}^*$:

$$\varepsilon_{ij}(\mathbf{V}^*) = \frac{1}{2}\left[\frac{\partial V_i^*}{\partial x_j} + \frac{\partial V_j^*}{\partial x_i}\right]. \tag{5.91}$$

It is important to introduce the bending moment $M(x)$, the shearing force $T(x)$ and the 1D-density of loading forces $f(x)$ defined by:

$$M(x) = -\int_{S(x)} y\sigma_{11}\,dS(x), \quad T(x) = \int_{S(x)} \sigma_{12}\,dS(x), \quad f(x) = \int_{S(x)} f_2\,dS(x), \tag{5.92}$$

where $S(x)$ denotes the section with abscissa x.

2) Without justifying the convergence of integrals, show that the principle of virtual works lead to the following formal variational formulation (**EVP**):

$$(\textbf{EVP}) \quad \begin{bmatrix} \text{Find } (M,T) \text{ defined from } [0,L] \times [0,L] \text{ to } \mathbf{R}, \text{ solution to:} \\[4pt] \int_0^L (\dot{M}+T)\,\Omega^*\,dx + \int_0^L (\dot{T}-f)\,v^*\,dx - M(L)\Omega^*(L) = 0, \\[4pt] \forall(v^*,\Omega^*)/v^*(0) = v^*(L) = \Omega^*(0) = 0. \end{bmatrix} \tag{5.93}$$

3) The behaviour laws of the constituent material of the beam are introduced:

$$M(x) = EI\dot{\Omega}(x) \quad \text{and} \quad T(x) = \mu S(\dot{v}(x) - \Omega(x)), \tag{5.94}$$

where I denotes the beam's moment of inertia in relation to the axis $(O;\mathbf{X}_3)$ and μ the shear stiffness coefficient.

Assuming that the force density f belongs to $L^2(0,L)$, show that the continuous problem **(CP)** composed of the equilibrium equations of the beam is expressed as:

(CP)

$$\left[\begin{array}{l} \text{Find } (v,\Omega) \text{ belonging to } H^2(0,L) \times H^2(0,L), \text{ solution of:} \\[2mm] EI\ddot{\Omega}(x) + \mu S[\dot{v}(x) - \Omega(x)] = 0, \quad \forall x \in]0,L[, \qquad (5.95) \\[2mm] \mu S[\ddot{v} - \dot{\Omega}] = f(x), \quad \forall x \in]0,L[, \qquad (5.96) \\[2mm] v(0) = v(L) = 0 \quad \text{and} \quad \Omega(0) = \dot{\Omega}(L) = 0. \qquad (5.97) \end{array}\right.$$

4) Introduce of the product space V, defined by:

$$V \equiv V_1 \times V_2 \equiv H^1_*(0,L) \times H^1_{**}(0,L),$$

where:

$$H^1_*(0,L) = \{\omega/\omega^{(k)} \in L^2(0,L), (k=0,1), \omega(0) = 0\}, \qquad (5.98)$$

and

$$H^1_{**}(0,L) = \{h/h^{(k)} \in L^2(0,L), (k=0,1), h(0) = h(L) = 0\}. \qquad (5.99)$$

Show that a moving variational formulation **(VP)** can be expressed as:

(VP)

$$\left\{\begin{array}{l} \text{Find } (\Omega,v) \text{ belonging to } V, \text{ solution of:} \\[2mm] EI\displaystyle\int_0^L \dot{\Omega}\dot{\omega}\,dx + \mu S\int_0^L (\dot{v} - \Omega)(\dot{h} - \omega)\,dx = \int_0^L fh\,dx, \qquad (5.100) \\[2mm] \forall (\omega,h) \in V. \end{array}\right.$$

5) Moreover, show that there is a minimization problem **(MP)** equivalent to the variational formulation **(VP)** defined as:

(MP)

$$\left\{\begin{array}{l} \text{Find } (v,\Omega) \in V \text{ solution of:} \\[2mm] J(v,\Omega) = \displaystyle\min_{(h,\omega)\in V} J(h,\omega), \text{ where:} \qquad (5.101) \\[2mm] J(h,\omega) = \dfrac{EI}{2}\displaystyle\int_0^L \dot{\omega}^2\,dx + \dfrac{\mu S}{2}\int_0^L [\dot{h} - \omega]^2\,dx + \int_0^L fh\,dx. \end{array}\right.$$

▶ **Numerical Part**

This part is dedicated to the approximation by finite elements $\mathbf{P_1}$ of bending elastic beams, modeled by the Timoshenko theory.

Thus, let the regular mesh of the interval $[0,L]$, having a constant step h, be such as:

$$\left\{\begin{array}{l} x_0 = 0, \ x_{N+1} = L, \\[2mm] x_{i+1} = x_i + h, \ i = 0 \text{ to } N. \end{array}\right. \qquad (5.102)$$

Moreover, the approximation spaces $(\tilde{V}_1, \tilde{V}_2)$ are introduced and defined by:

$$\tilde{V}_1 = \{\tilde{h}/\tilde{h} \in C^0([0,L]), \tilde{h}|_{[x_i,x_{i+1}]} \in P_1, \ \tilde{h}(0) = \tilde{h}(L) = 0\}, \quad (5.103)$$

$$\tilde{V}_2 = \{\tilde{\omega}/\tilde{\omega} \in C^0([0,L]), \tilde{\omega} \in P_1, \ \tilde{\omega}(0) = 0\}, \quad (5.104)$$

where $P_1 \equiv P_1([x_i, xi_{+1}])$ refers to the space of polynomials defined on $[x_i, x_{i+1}]$, having a degree less than or equal to one.

6) What is the dimension of spaces $\tilde{V}_1$ and $\tilde{V}_2$?

7) Let $(\tilde{v}, \tilde{\Omega})$ belong to $\tilde{V} \equiv \tilde{V}_1 \times \tilde{V}_2$ as defined by:

$$\tilde{v} = \sum_{j=1}^{N} \tilde{v}_j \varphi_j, \ \tilde{\Omega} = \sum_{k=1}^{N+1} \tilde{\Omega}_k \varphi_k, \quad (5.105)$$

where φ_i is the functions of the canonical basis associated with the finite elements $\mathbf{P}_1$ satisfying: $\varphi_i(x_j) = \delta_{ij}$.

Let $(\widetilde{\mathbf{MP}})$, the approximate minimization problem associated with problem $(\mathbf{MP})$ be defined by:

$$(\widetilde{\mathbf{PM}}) \quad \begin{bmatrix} \text{Find } (\tilde{v}, \tilde{\Omega}) \in \tilde{V}, \text{ solution to:} \\ J(\tilde{v}, \tilde{\Omega}) = \underset{(\tilde{h}, \tilde{\omega}) \in \tilde{V}}{\text{Min}} J(\tilde{h}, \tilde{\omega}), \end{bmatrix} \quad (5.106)$$

where J is the potential energy defined by (5.101).

– Show that a necessary condition allowing $(\tilde{v}, \tilde{\Omega})$ to be a solution of the approximate minimization problem $(\widetilde{\mathbf{MP}})$ is written in the following form:

Find the two sequences having components $(\tilde{v}_j)_{j=1,N}$ and $(\tilde{\Omega}_k)_{k=1,N+1}$, defining the approximation belonging to $\tilde{V}$, solution to the global linear system:

$$(\widetilde{\mathbf{MP}}) \quad \begin{bmatrix} A_{ij}^{(1,1)}\tilde{v}_j + A_{ij}^{(2,1)}\tilde{\Omega}_j = b_i^{(1)}, & (i = 1 \text{ to } N), & (5.107) \\ A_{ij}^{(1,2)}\tilde{v}_j + A_{ij}^{(2,2)}\tilde{\Omega}_j = b_i^{(2)}, & (i = 1 \text{ to } N+1), \\ (\forall j \, / \, Supp \ \varphi_i \cap Supp \ \varphi_j \neq \varnothing), & & (5.108) \end{bmatrix}$$

where:

$$A_{ij}^{(1,1)} = \mu S \int_0^L \dot{\varphi}_i \dot{\varphi}_j \, dx,$$

$$A_{ij}^{(1,2)} = -\mu S \int_0^L \dot{\varphi}_i \varphi_j \, dx, \quad (5.109)$$

$$A_{ij}^{(2,1)} = -\mu S \int_0^L \varphi_i \dot{\varphi}_j \, dx,$$

$$A_{ij}^{(2,2)} = EI \int_0^L \dot{\varphi}_i \dot{\varphi}_j \, dx + \mu S \int_0^L \varphi_i \varphi_j \, dx \,, \tag{5.110}$$

$$b_i^{(1)} = - \int_0^L f \varphi_i \, dx \,, \tag{5.111}$$

$$b_i^{(2)} = 0 \,. \tag{5.112}$$

Moreover, it would have been agreed to adopt the Einstein Summation Convention (or summation of repeated indices convention):

$$A_{ij}^{(1,1)} \tilde{v}_j \equiv \sum_j A_{ij}^{(1,1)} \tilde{v}_j \,. \tag{5.113}$$

8) For each of the characteristic basis functions $(\varphi_i)_{1,N}$ of each node strictly interior at $[0, L]$, write the corresponding nodal equation associated with the discrete equs. (5.107) and (5.108).

9) Show that the approximation of the system of the differential equs. (5.95)–(5.96) is then found again by the finite differences method.

10) Write the nodal equation associated with the basis function φ_{N+1}, characteristic of node x_{N+1}, in relation to the discrete equation (5.108).

11) Find the finite differences scheme that discretises the Neumann condition of equ. (5.97) bearing on the first derivative of Ω.

12) Show that the elementary stiffness matrix governing the contribution of the local element $[x_i, x_{i+1}]$ and associated with the system (5.107)–(5.108) is written as:

$$A^{(i+1)} = \begin{bmatrix} \alpha & \beta & -\alpha & \beta \\ \beta & \delta & -\beta & \gamma \\ -\alpha & -\beta & \alpha & -\beta \\ \beta & \gamma & -\beta & \delta \end{bmatrix}, \tag{5.114}$$

where the following was noted:

$$\alpha = \frac{\mu S}{h}, \quad \beta = \frac{\mu S}{2}, \quad \gamma = -\frac{EI}{h}, \quad \delta = \left[\frac{EI}{h} + \frac{\mu Sh}{2}\right]. \tag{5.115}$$

13) Then, proceed to the assembly of the discrete system having an approximation $\widetilde{(MP)}$ defined by (5.107)–(5.108).

Solution

▶ **Displacements Variational Formulation – Principle of Virtual Works – Theoretical Part**

A.1) The form of the virtual displacements field $\mathbf{V}^*$ defined by (5.88) implies that the strain tensor $\varepsilon(\mathbf{V}^*)$ (cf. (5.91)) associated with $\mathbf{V}^*$ is written as:

$$\varepsilon(\mathbf{V}^*) = \varepsilon_{11}^* \mathbf{X}_1 \otimes \mathbf{X}_1 + \varepsilon_{12}^* [\mathbf{X}_1 \otimes \mathbf{X}_2 + \mathbf{X}_1 \otimes \mathbf{X}_1] , \qquad (5.116)$$

where:

$$\varepsilon_{11}^* = -y\dot{\Omega}^*(x) \quad \text{and} \quad \varepsilon_{12}^* = \frac{1}{2}[\dot{v}^*(x) - \Omega^*(x)] . \qquad (5.117)$$

The notation of the tensor product $\mathbf{X}_i \otimes \mathbf{X}_j$ resulting in the presence of a non-zero component ε_{ij}^* within the strain tensor $\varepsilon(\mathbf{U}^*)$ would have been used.

The virtual work of the internal forces T_{Int}^* is then evaluated as follows:

$$
\begin{aligned}
T_{Int}^* &= -\int_\Omega \sigma_{ij}\varepsilon_{ij}(\mathbf{V}^*)\,dv \\
&= -\left[\int_\Omega \sigma_{11}\varepsilon_{11}(\mathbf{V}^*)\,dv + 2\int_\Omega \sigma_{12}\varepsilon_{12}(\mathbf{V}^*)\,dv\right] \\
&= -\int_0^L \left(-\iint_{S(x)} y\sigma_{11}\,dS(x)\right)\dot{\Omega}^*(x)\,dx \\
&\quad - \int_0^L \left(\iint_{S(x)} \sigma_{12}\,dS(x)\right)[\dot{v}^*(x) - \Omega^*(x)]\,dx \\
&= -\int_0^L M(x)\dot{\Omega}^*(x)\,dx - \int_0^L T(x)[\dot{v}^*(x) - \Omega^*(x)]\,dx , \qquad (5.118)
\end{aligned}
$$

where the definitions of the bending moment $M(x)$ and of the shearing force $T(x)$ defined by (5.92) were used.

Similarily, the virtual work of the external forces is obtained from:

$$
\begin{aligned}
T_{Ext}^* &= \int_\Omega \mathbf{f} \cdot \mathbf{V}^*\,dv = -\int_\Omega f_2 v^*(x)\,dx \\
&= -\int_0^L \left(\iint_{S(x)} f_2\,dS(x)\right) v^*(x)\,dx = -\int_0^L f(x)v^*(x)\,dx . \qquad (5.119)
\end{aligned}
$$

A.2) The application of the principle of virtual works is then written within the framework of an elastostatic problem:

$$T_{Int}^* + T_{Ext}^* = 0, \quad \forall V^* \in \mathfrak{V}^* . \qquad (5.120)$$

Thus, using expressions (5.118) and (5.119), the following is obtained:

$$\int_0^L M\dot{\Omega}^* \, dx + \int_0^L T[\dot{v}^* - \Omega^*] \, dx + \int_0^L fv^* \, dx = 0 ,$$

$$\forall (v^*, \Omega^*) / v^*(0) = v^*(L) = \Omega^*(0) = 0 . \tag{5.121}$$

The use of integrations by parts produces a transformation of the variational formulation (5.121) as follows:

$$\int_0^L \dot{M}\Omega^* \, dx - [M\Omega^*]_0^L + \int_0^L \dot{T}v^* \, dx - [Tv^*]_0^L + \int_0^L T\Omega^* \, dx - \int_0^L fv^* \, dx = 0 ,$$

$$\forall (v^*, \Omega^*) / v^*(0) = v^*(L) = \Omega^*(0) = 0 . \tag{5.122}$$

The boundary conditions (5.89) of the virtual fields V^* lead to the variational formulation (**EVP**):

$$(\mathbf{EVP}) \left[\begin{array}{l} \text{Find } (M,T) \text{ defined from } [0,L] \times [0,L] \text{ to } \mathbf{R}, \text{ solution to:} \\[2mm] \int_0^L (\dot{M} + T) \, \Omega^* \, dx + \int_0^L (\dot{T} - f) \, v^* \, dx - M(L)\Omega^*(L) = 0 , \qquad (5.123) \\[2mm] \forall (v^*, \Omega^*) / v^*(0) = v^*(L) = \Omega^*(0) = 0 . \end{array} \right.$$

A.3) The continuous problem (**CP**) is obtained using the density arguments just as they were applied on several occasions.

To achieve this, first of all the variational equation (5.123) is transformed by replacing the bending moment M and the shearing force T by the kinematic variables (v, Ω), by applying the behaviour laws (5.94).

The variational problem (**EVP**) is then written as:

$$(\mathbf{EVP}) \left[\begin{array}{l} \text{Find } (v, \Omega) \text{ defined by } [0,L] \times [0,L] \text{ having values in } \mathbf{R}, \text{ solution to:} \\[2mm] \int_0^L \left[EI\ddot{\Omega} + \mu S(\dot{v} - \Omega) \right] \Omega^* \, dx + \int_0^L \left[\mu S(\dot{v} - \dot{\Omega}) - f \right] v^* \, dx \qquad (5.124) \\[2mm] -EI\dot{\Omega}(L)\Omega^*(L) = 0, \ \forall (v^*, \Omega^*) / v^*(0) = v^*(L) = \Omega^*(0) = 0 . \end{array} \right.$$

A reasonable functional framework is now defined to give a sense to the writing up of problem (5.124).

To achieve this, suppose that v and v^* are elements of $H_{**}^2(0,L)$ – defined in the same way as (5.99) provided that H^1 is replaced by H^2 (it is only necessary to consider the virtual fields v^* in $L^2(0,L)$), whereas the rotation of the section Ω is looked for in the space $H_*^2(0,L)$ defined according to the principle similar to (5.98).

Under these conditions, it is easily noticed that the integral equation (5.124) makes sense, and this is essentially due to the Cauchy-Schwartz inequality.

In fact, it is only necessary to note that the quantities $[EI\ddot{\Omega} + \mu S(\dot{v} - \Omega)]$ and $[\mu S(\dot{v} - \dot{\Omega}) - f]$ are really elements of $L^2(0,L)$, (f itself being a distribution of given forces belonging to $L^2(0,L)$).

Then, consider the particular case where equ. (5.124) is satisfied for the virtual fields Ω^* which are identically zero.

The particular formulation of (5.124) is then written as:

Find (v, Ω) belonging to $H^2_{**}(0,L) \times H^2_*(0,L)$ solution to:

$$\int_0^L \left(\mu S(\dot{v} - \dot{\Omega}) - f\right) v^* \, dx, \quad \forall v^* \in H^2_{**}(0,L) . \tag{5.125}$$

The differential equation (5.96) is obtained by using the density method.

Among the functions v^* of $H^2_{**}(0,L)$, consider those belonging to $\mathscr{D}(0,L)$ (since $\mathscr{D}(0,L) \subset H^2_{**}(0,L)$) and use the fact that $\mathscr{D}(0,L)$ is dense in $L^2(0,L)$.

A complete demonstration of the density method may be consulted within the framework of the problem of a beam subjected to traction [5.1.1].

Finally, the following is obtained:

$$\mu S(\ddot{v} - \dot{\Omega})(x) - f(x) = 0, \quad \forall x \in]0,L[. \tag{5.126}$$

A similar reasoning leads to the obtention of differential equation (5.95) by the same arguments of density and by considering this time as a second particular case of the equ. (5.124) that of the functions v^* which are identically zero.

At last, having the two differential equations of the continuous problem **(CP)**, the equ. (5.24) is degenerate and is written as:

$$-EI\dot{\Omega}(L)\Omega^*(L) = 0, \quad \forall \Omega^* / \Omega^*(0) = 0 . \tag{5.127}$$

It is immediately inferred that: $\dot{\Omega}(L) = 0$.

▶ **Summary:**

The solution (v, Ω) of the variational equation (5.124) was considered in the product space $H^2_{**}(0,L) \times H^2_*(0,L)$, satisfying *de facto* the boundary conditions (5.95) when $x = 0$ and when $x = L$.

Moreover, it is proved by the density method that v and Ω are solutions of the differential system (5.95)–(5.96).

In other words, (v, Ω) is a solution to the continuous problem **(CP)** defined by (5.95)–(5.97).

A.4) Let (v, Ω) be a solution to the continuous problem **(CP)** defined by (5.95)–(5.97). Moreover, let there be a pair of test functions (h, ω).

Then, the differential equation (5.95) of the problem (**CP**) is multiplied by the test function h and the differential equation (5.96) by the second test function ω.

Then, each equation is integrated within $[0, L]$ to obtain:

$$\left[EI \int_0^L \ddot{\Omega} \omega \, dx = -\mu S \int_0^L (\dot{v} - \Omega) \omega \, dx, \quad \forall \omega \in V_1, \right. \tag{5.128}$$

$$\left. \mu S \int_0^L (\ddot{v} - \dot{\Omega}) h \, dx = \int_0^L f h \, dx, \quad \forall h \in V_2, \right. \tag{5.129}$$

where $(V_1 \times V_2)$ represents a pair of spaces of the test functions (h, ω) that will be defined later.

Then, an integration by parts of the integral equations (5.128)–(5.129) is performed to obtain:

$$\left[-EI \int_0^L \dot{\Omega} \dot{\omega} \, dx + EI \left[\dot{\Omega} \omega \right]_0^L = -\mu S \int_0^L (\dot{v} - \Omega) \omega \, dx, \quad \forall \omega \in V_1, \right. \tag{5.130}$$

$$\left. -\mu S \int_0^L \dot{v} h \, dx + \mu S \left[\dot{v} h \right]_0^L - \mu S \int_0^L \dot{\Omega} h \, dx = \int_0^L f h \, dx, \quad \forall h \in V_2. \right. \tag{5.131}$$

Thus, the unique possibility to take into account the Neumann boundary conditions bearing on Ω appears in the formulation (5.130)–(5.131): $\dot{\Omega}(L) = 0$.

Other Dirichlet-type boundary conditions bearing on v and Ω must be kept in the future variational formulation, by using the test functions h and ω.

To achieve this, the pair of test functions (ω, h) is made to belong to the functional product space $V \equiv V_1 \times V_2 \equiv H^1_*(0, L) \times H^1_{**}(0, L)$ with:

$$H^1_*(0, L) = \left\{ \omega / \omega^{(k)} \in L^2(0, L), (k = 0, 1), \omega(0) = 0 \right\}, \tag{5.132}$$

$$H^1_{**}(0, L) = \left\{ h / h^{(k)} \in L^2(0, L), (k = 0, 1), h(0) = h(L) = 0 \right\}. \tag{5.133}$$

The immediate consequence of this choice of functional spaces is the disappearance of the terms entirely integrated in the system (5.130)–(5.131) which is then written as:

Find (ω, v) belonging to V solution to:

$$\left[-EI \int_0^L \dot{\Omega} \dot{\omega} \, dx = -\mu S \int_0^L (\dot{v} - \Omega) \omega \, dx, \quad \forall \omega \in V_1, \right. \tag{5.134}$$

$$\left. -\mu S \int_0^L \dot{v} h \, dx - \mu S \int_0^L \dot{\Omega} h = \int_0^L f h \, dx, \quad \forall h \in V_2. \right. \tag{5.135}$$

Moreover, it will be noticed that the pair solution (Ω, v) has lost much of its regularity during the change from strong formulation of the continuous problem (**CP**) to the weak or variational formulation (**VP**).

In fact, the functions belonging to $V_1 \times V_2$ are sufficiently regular to guarantee the convergence of the integrals appearing in the variational formulation (**VP**). In order to obtain the variational formulation (**VP**) defined by (5.100), it is only necessary to sum the two equs. (5.134) and (5.135).

A.5) The variational formulation (**VP**) defined by (5.100) presents the advantage of having a bilinear form $a[(.,.),(.,.)]$ and a linear form $L[(.,.)]$ defined by:

$$a[(\Omega,v),(\omega,h)] = EI \int_0^L \dot{\Omega}\dot{\omega}\,dx + \mu S \int_0^L (\dot{v}-\Omega)(\dot{h}-\omega)\,dx, \qquad (5.136)$$

$$L[(\omega,h)] = \int_0^L fh\,dx. \qquad (5.137)$$

It is then observed that the two forms a and L possess the whole properties including the symmetry of the bilinear form a in particular – D. Euvrard [4] may be consulted to make a list of the whole properties – so as to obtain, by equivalence, the minimization problem (**MP**) defined by:

$$(\mathbf{MP}) \begin{cases} \text{Find } (v,\Omega) \in V \text{ solution of:} \\[2mm] J(v,\Omega) = \underset{(h,\omega)\in V}{\text{Min}}\ J(h,\omega) \text{ where:} \\[2mm] J(h,\omega) = \dfrac{1}{2}a[(h,\omega),(h,\omega)] - L[(h,\omega)]. \end{cases} \qquad (5.138)$$

▶ **Numerical Part**

A.6) The estimation of the dimension of spaces $\tilde{V}_1$ and $\tilde{V}_2$ is performed according to the same procedure. Only the additional degree of freedom when $x = L$ is to be considered in the functions belonging to $\tilde{V}_1$, which provides one unit of difference compared to the dimension of space $\tilde{V}_2$.

Thus, the functions belonging to space $\tilde{V}_1$ are exactly defined by the N values of the interior nodes $(x_i)_{i=1,N}$ defined by the mesh (5.102). To ascertain that, it is only necessary to proceed to the visualization of such functions (cf. Fig. 5.4).

In fact, the functions of the approximation space $\tilde{V}_1$ are pecked lines formed by affine functions per mesh $[x_i, x_{i+1}]$, whose interior nodes x_i constitute the points of continuity between two adjacent meshes.

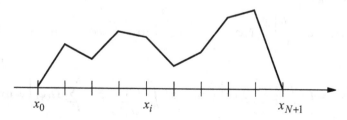

Fig. 5.4 Profile of a Piecewise Affine Function

Moreover, concerning the functions belonging to space $\tilde{V}_1$, it is advisable to add the two stresses which require the functions of this space to be zero when $x = 0$ and when $x = L$.

That is why only values at the N nodes strictly interior at the interval $[0, L]$ display a degree of freedom for any function $\tilde{v}$ of $\tilde{V}_1$. Changing of one of these N values immediately results in the modification of element $\tilde{v}$ of $\tilde{V}$ considered in another function of $\tilde{V}_1$.

Thus, without any formal demonstration, it is observed that understanding a function $\tilde{v}$ of $\tilde{V}_1$ is equivalent to understanding a vector $\mathbf{R}^N$ constituted by the N values of $(\tilde{v}_1, \dots, \tilde{v}_N)$ at the discretisation points $(x_1, \dots, x_N)$.

In conclusion, given the additional degree of freedom when $x = L$, the dimension of $\tilde{V}_1$ is equal to N whereas that of $\tilde{V}_2$ is equal to $(N + 1)$.

A.7) The approximate formulation $\widetilde{(\mathbf{MP})}$ associated with the minimization problem $(\mathbf{MP})$ is obtained by evaluating functional J defined by (5.101) at point $(\tilde{h}, \tilde{v})$:

$$J(\tilde{h}, \tilde{\omega}) = \frac{EI}{2} \int_0^L \dot{\tilde{\omega}}^2 \, dx + \frac{\mu S}{2} \int_0^L [\dot{\tilde{h}} - \tilde{\omega}]^2 \, dx + \int_0^L f \tilde{h} \, dx, \qquad (5.139)$$

where $(\tilde{h}, \tilde{v})$ belongs to $\tilde{V}_1 \times \tilde{V}_2$.

Thus, given the finite dimension of spaces V_1 and V_2, $\tilde{h}$ and $\tilde{v}$ can be described in the form of developments which are similar to (5.105).

In that case, $J(\tilde{h}, \tilde{v})$ is evaluated as follows:

$$J(\tilde{h}, \tilde{\omega}) = \frac{EI}{2} \int_0^L \left[\sum_{k=1}^{N+1} \tilde{\omega}_k \dot{\varphi}_k \right]^2 dx + \frac{\mu S}{2} \int_0^L \left[\sum_{j=1}^{N} \tilde{h}_j \dot{\varphi}_j - \sum_{k=1}^{N+1} \tilde{\omega}_k \varphi_k \right]^2 dx$$

$$+ \int_0^L f \sum_{j=1}^{N} \tilde{h}_j \varphi_j \, dx . \qquad (5.140)$$

A necessary condition enabling $(\tilde{v}, \tilde{\Omega})$ to constitute the minimum of function J is then written as:

$$\frac{\partial J}{\partial \tilde{h}_i}(\tilde{v}, \tilde{\Omega}) = \frac{\partial J}{\partial \tilde{\omega}_l}(\tilde{v}, \tilde{\Omega}) = 0, \quad \forall (i, l) \in \{1, \dots, N\} \times \{1, \cdots, N+1\} . \qquad (5.141)$$

The following is then obtained:

$$\frac{\partial J}{\partial \tilde{h}_i}(\tilde{v}, \tilde{\Omega}) = \mu S \int_0^L \left[\sum_{j=1}^{N} \tilde{v}_j \dot{\varphi}_j - \sum_{k=1}^{N+1} \tilde{\Omega}_k \varphi_k \right] \dot{\varphi}_i \, dx + \int_0^L f \varphi_i \, dx = 0, \quad (\forall i = 1, N) .$$

$$\qquad (5.142)$$

$$\frac{\partial J}{\partial \tilde{\omega}_l}(\tilde{v}, \tilde{\Omega}) = EI \int_0^L \left[\sum_{k=1}^{N+1} \tilde{\Omega}_k \dot{\varphi}_k \right] \dot{\varphi}_l \, dx - \mu S \int_0^L \left[\sum_{j=1}^{N} \tilde{v}_j \dot{\varphi}_j - \sum_{k=1}^{N+1} \tilde{\Omega}_k \varphi_k \right] \varphi_l \, dx$$

$$= 0, (\forall l = 1, N+1) . \qquad (5.143)$$

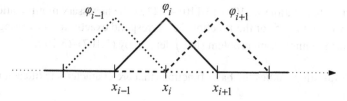

Fig. 5.5 Basis Functions φ_{i-1}, φ_i and φ_{i+1}

Then, it is noticed that the support of a basis function φ_i consists of two meshes $[x_{i-1},x_i]$ and $[x_i,x_{i+1}]$, (cf. Fig. 5.5).

From now on, only the basis functions φ_{i-1}, φ_i and φ_{i+1} have a support whose intersection with that of the function φ_i is non-vacuous.

That is why the finite sums intervening in equs. (5.142) and (5.143) are degenerate in the following way:

$$\left[\frac{\partial J}{\partial \tilde{h}_i}(\tilde{v},\tilde{\Omega}) = 0\right] \Leftrightarrow$$

$$\mu S\left[\left(\int_0^L \dot{\varphi}_{i-1}\dot{\varphi}_i\right)\tilde{v}_{i-1} - \left(\int_0^L \varphi_{i-1}\dot{\varphi}_i\right)\tilde{\Omega}_{i-1}\right] + \cdots$$

$$\mu S\left[\left(\int_0^L \dot{\varphi}_i^2\right)\tilde{v}_i - \left(\int_0^L \varphi_i\dot{\varphi}_i\right)\tilde{\Omega}_i\right] + \cdots$$

$$\mu S\left[\left(\int_0^L \dot{\varphi}_{i+1}\dot{\varphi}_i\right)\tilde{v}_{i+1} - \left(\int_0^L \varphi_{i+1}\dot{\varphi}_i\right)\tilde{\Omega}_{i+1}\right] = -\int_0^L f\varphi_i .$$

$$(5.144)$$

Likewise:

$$\left[\frac{\partial J}{\partial \tilde{\omega}_l}(\tilde{v},\tilde{\Omega}) = 0\right] \Leftrightarrow$$

$$EI\left[\left(\int_0^L \dot{\varphi}_{l-1}\dot{\varphi}_l\right)\tilde{\Omega}_{l-1}\right.$$

$$\left. + \left(\int_0^L \dot{\varphi}_l^2\right)\tilde{\Omega}_l + \left(\int_0^L \dot{\varphi}_{l+1}\dot{\varphi}_l\right)\tilde{\Omega}_{l+1}\right]$$

$$- \mu S\left[\left(\int_0^L \dot{\varphi}_{l-1}\varphi_l\right)\tilde{v}_{l-1} - \left(\int_0^L \varphi_{l-1}\varphi_l\right)\tilde{\Omega}_{l-1}\right]$$

$$- \mu S\left[\left(\int_0^L \dot{\varphi}_l\varphi_l\right)\tilde{v}_l - \left(\int_0^L \varphi_l^2\right)\tilde{\Omega}_l\right]$$

$$- \mu S\left[\left(\int_0^L \dot{\varphi}_{l+1}\varphi_l\right)\tilde{v}_{l+1} - \left(\int_0^L \varphi_{l+1}\varphi_l\right)\tilde{\Omega}_{l+1}\right] = 0 . \quad (5.145)$$

By reverting to notations (5.109), (5.110)–(5.112), the necessary minimization conditions (5.144)–(5.145) of the functional J correspond exactly to the writing of the approximate minimization problem $(\widetilde{MP})$ defined by (5.107)–(5.108).

A.8) This question studies the basis functions that are characteristic of each interior node x_i, $(i = 1, N)$.

The equation system is then re-written to define the approximate formulation $(\widetilde{MP})$ by using the elementary properties of the basis functions φ_i.

Furthermore, the trapezium quadrature formula is used to evaluate the integrals that occur in equs. (5.144)–(5.145).

It then becomes:

$(\forall i = 1, N), \quad (\forall l = 1, N):$

$$\left[\frac{\partial J}{\partial \tilde{h}_i}(\tilde{v}, \tilde{\Omega}) = 0\right] \Leftrightarrow \mu S \left[-\frac{\tilde{v}_{i-1} - 2\tilde{v}_i + \tilde{v}_{i+1}}{h^2} + \frac{\tilde{\Omega}_{i+1} - \tilde{\Omega}_{i-1}}{2h}\right] = -f_i, \quad (5.146)$$

$$\left[\frac{\partial J}{\partial \tilde{\omega}_l}(\tilde{v}, \tilde{\Omega}) = 0\right] \Leftrightarrow EI \left[\frac{\tilde{\Omega}_{l-1} - 2\tilde{\Omega}_l + \tilde{\Omega}_{l+1}}{h^2}\right] + \mu S \left[\frac{\tilde{v}_{l+1} - \tilde{v}_{l-1}}{2h} - \tilde{\Omega}_l\right] = 0 .$$
$$(5.147)$$

A.9) To work out an approximation by finite differences of the second order of differential equations (5.95)–(5.96) from nodal equations (5.146)–(5.147), the following different identities, as demonstrated on several occasions are observed:

$$\frac{v_{i-1} - 2v_i + v_{i+1}}{h^2} = \ddot{v}(x_i) + O(h^2), \qquad (5.148)$$

$$\frac{\Omega_{i+1} - \Omega_{i-1}}{2h} = \dot{\Omega}(x_i) + O(h^2). \qquad (5.149)$$

The nodal equations (5.146)–(5.147) are respectively written as:

$$\mu S \left[\ddot{v}(x_i) - \dot{\Omega}(x_i)\right] = f(x_i), \quad (i = 1, N), \qquad (5.150)$$

$$EI\ddot{\Omega}(x_l) + \left[\dot{v}(x_l) - \tilde{\Omega}(x_l)\right] = 0, \quad (l = 1, N), \qquad (5.151)$$

to the nearest $O(h^2)$. Hence the equs. (5.150)–(5.151) strictly correspond to differential equations (5.95)–(5.96) of the continuous problem $(\mathbf{CP})$, provided that the approximation functions $(\tilde{v}, \tilde{\Omega})$ have been substituted by the solutions (v, Ω).

A.10) To write down the nodal equation which corresponds to the basis function φ_{N+1}, the final equation of system (5.143) is considered and is written in this particular case as:

$$\left[\frac{\partial J}{\partial \tilde{\omega}_{N+1}}(\tilde{v},\tilde{\Omega}) = 0 \right] \Leftrightarrow EI \left[\left(\int_0^L \dot{\varphi}_N \dot{\varphi}_{N+1} \right) \tilde{\Omega}_N + \left(\int_0^L \dot{\varphi}_{N+1}^2 \right) \tilde{\Omega}_{N+1} \right]$$

$$- \mu S \left[\left(\int_0^L \dot{\varphi}_N \varphi_{N+1} \right) \tilde{v}_N - \left(\int_0^L \varphi_N \varphi_{N+1} \right) \tilde{\Omega}_N \right]$$

$$- \mu S \left[\left(\int_0^L \dot{\varphi}_{N+1} \varphi_{N+1} \right) \tilde{v}_{N+1} - \left(\int_0^L \varphi_{N+1}^2 \right) \tilde{\Omega}_{N+1} \right] = 0 .$$

$$(5.152)$$

The trapezium quadrature formula, combined with the elementary properties of the basis functions φ_N and φ_{N+1} (cf. Fig. 5.6) is once more applied and the following is obtained:

$$EI \frac{\tilde{\Omega}_{N+1} - \tilde{\Omega}_N}{h} + \frac{\mu S}{2} \left[\tilde{v}_N - \tilde{v}_{N+1} \right] + \frac{\mu S h}{2} \tilde{\Omega}_{N+1} = 0 . \qquad (5.153)$$

A.11) This question involves a discretisation of the Neumann condition (5.97) by finite differences: $\dot{\Omega}(x_{N+1}) = 0$.

To achieve this, the regressive Taylor's expansion is written at the abscissa x_{N+1}:

$$\Omega(x_N) = \Omega(x_{N+1}) - h\dot{\Omega}(x_{N+1}) + \frac{h^2}{2}\ddot{\Omega}(x_{N+1}) + O(h^3) . \qquad (5.154)$$

Now, the second derivative $\ddot{\Omega}$ is determined at abscissa x_{N+1} by assuming that the first differential equation (5.95) of the continuous problem (**CP**) may be extended by continuity at this point.

It then becomes:

$$\Omega(x_N) = \Omega(x_{N+1}) + \frac{h^2}{2} \left[\frac{\mu S}{EI} \left(\Omega(x_{N+1}) - \dot{v}(x_{N+1}) \right) \right] + O(h^3). \qquad (5.155)$$

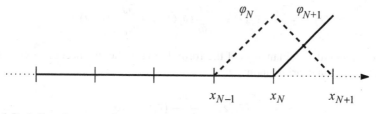

Fig. 5.6 Basis Functions φ_N and φ_{N+1}

To maintain a second-order approximation in h, it is only necessary to consider a first-order approximation of the first derivative $\dot{v}$ at the point x_{N+1}, the first derivative being weighted by a multiple of h^2.

Therefore, the common regressive finite difference is considered:

$$\dot{v}(x_{N+1}) = \frac{v(x_{N+1}) - v(x_N)}{h} + O(h) . \tag{5.156}$$

The expression $\dot{v}(x_{N+1})$ supplied by (5.156) is then replaced in the limited expansion (5.155):

$$EI\left[\Omega(x_N) - \Omega(x_{N+1})\right] - \frac{\mu S h^2}{2}\Omega(x_{N+1}) = \frac{\mu S h}{2}\left[v(x_{N+1}) - v(x_N)\right] + O(h^3) , \tag{5.157}$$

or even:

$$EI\left[\frac{\Omega(x_N) - \Omega(x_{N+1})}{h}\right] - \frac{\mu S h}{2}\Omega(x_{N+1}) - \frac{\mu S}{2}\left[v(x_{N+1}) - v(x_N)\right] = O(h^2) . \tag{5.158}$$

This last formulation helps to find the exact nodal equation corresponding to the basis function φ_{N+1} described in the final equation of the system (5.143).

To achieve this, it suffices to use the approximations in the equ. (5.158) by substituting the real values $(v_i, \Omega_i) \equiv (v, \Omega)(x_i)$, $(i = N, N+1)$ by their corresponding approximations $(\tilde{v}_i, \tilde{\Omega}_i)$ and by eliminating the residue of Landau $O(h^2)$.

A.12) To determine the elementary matrix $A^{(i+1)}$, which expresses the local element contribution $[x_i, x_{i+1}]$ in the global system of the approximate minimisation problem $\widetilde{(MP)}$ defined in (5.107) – (5.108), the minimisation equations causing application of the geometrical mesh are written down.

Yet, only the basis functions φ_i and φ_{i+1} have part of their support that intercept the interval $[x_i, x_{i+1}]$.

This explains why only the minimisation equations, which are responsible for the occurrence of the basis functions φ_i and φ_{i+1} along the interval $[x_{i,i+1}]$ are written down.

$$\frac{\partial J}{\partial \tilde{h}_i}(\tilde{v}, \tilde{\Omega}) = \frac{\partial J}{\partial \tilde{h}_{i+1}}(\tilde{v}, \tilde{\Omega}) = \frac{\partial J}{\partial \tilde{\omega}_i}(\tilde{v}, \tilde{\Omega}) = \frac{\partial J}{\partial \tilde{\omega}_{i+1}}(\tilde{v}, \tilde{\Omega}) = 0 . \tag{5.159}$$

Below is a detailed explanation of the formulation of the two necessary conditions for minimisation bearing upon:

$$\frac{\partial J}{\partial \tilde{h}_i}(\tilde{v}, \tilde{\Omega}) = \frac{\partial J}{\partial \tilde{h}_{i+1}}(\tilde{v}, \tilde{\Omega}) = 0 . \tag{5.160}$$

The following is obtained:

$$\left[\frac{\partial J}{\partial \tilde{h}_i}(\tilde{v},\tilde{\Omega}) = 0\right] \Leftrightarrow \cdots$$

$$\tilde{v}_{i-1}\left[\mu S \int_{x_{i-1}}^{x_i} \dot{\varphi}_{i-1}\dot{\varphi}_i\right] + \cdots$$

$$\tilde{v}_i\left[\mu S \int_{x_{i-1}}^{x_i} \dot{\varphi}_i^2 + \mu S \int_{x_i}^{x_{i+1}} \dot{\varphi}_i^2\right] + \cdots$$

$$\tilde{v}_{i+1}\left[\mu S \int_{x_i}^{x_{i+1}} \dot{\varphi}_{i+1}\dot{\varphi}_i\right] + \cdots$$

$$\tilde{\Omega}_{i-1}\left[-\mu S \int_{x_{i-1}}^{x_i} \varphi_{i-1}\dot{\varphi}_i\right] + \cdots$$

$$\tilde{\Omega}_i\left[-\mu S \int_{x_{i-1}}^{x_i} \varphi_i\dot{\varphi}_i - \mu S \int_{x_i}^{x_{i+1}} \varphi_i\dot{\varphi}_i\right] + \cdots$$

$$\tilde{\Omega}_{i+1}\left[-\mu S \int_{x_i}^{x_{i+1}} \varphi_{i+1}\dot{\varphi}_i\right] + \cdots$$

$$= -\left[\int_{x_{i-1}}^{x_i} f\varphi_i + \int_{x_i}^{x_{i+1}} f\varphi_i\right]. \qquad (5.161)$$

Then,

$$\left[\frac{\partial J}{\partial \tilde{h}_{i+1}}(\tilde{v},\tilde{\Omega}) = 0\right] \Leftrightarrow \cdots$$

$$\tilde{v}_i\left[\mu S \int_{x_i}^{x_{i+1}} \dot{\varphi}_i\dot{\varphi}_{i+1}\right] + \cdots$$

$$\tilde{v}_{i+1}\left[\mu S \int_{x_i}^{x_{i+1}} \dot{\varphi}_{i+1}^2 + \mu S \int_{x_{i+1}}^{x_{i+2}} \dot{\varphi}_{i+1}^2\right] + \cdots$$

$$\tilde{v}_{i+2}+\left[\mu S \int_{x_{i+1}}^{x_{i+2}} \dot{\varphi}_{i+2}\dot{\varphi}_{i+1}\right] + \cdots$$

$$\tilde{\Omega}_i\left[-\mu S \int_{x_i}^{x_{i+1}} \varphi_i\dot{\varphi}_{i+1}\right] + \cdots$$

$$\tilde{\Omega}_{i+1}\left[-\mu S \int_{x_i}^{x_{i+1}} \varphi_{i+1}\dot{\varphi}_{i+1} - \mu S \int_{x_{i+1}}^{x_{i+2}} \varphi_{i+1}\dot{\varphi}_{i+1}\right] + \cdots$$

$$\tilde{\Omega}_{i+2}\left[-\mu S \int_{x_{i+1}}^{x_{i+2}} \varphi_{i+2}\dot{\varphi}_{i+1}\right] + \cdots$$

$$= -\left[\int_{x_i}^{x_{i+1}} f\varphi_{i+1} + \int_{x_{i+1}}^{x_{i+2}} f\varphi_{i+1}\right]. \qquad (5.162)$$

Likewise, the two other conditions for minimisation can at present be explained in detail:

$$\frac{\partial J}{\partial \tilde{\omega}_i}(\tilde{v},\tilde{\Omega}) = \frac{\partial J}{\partial \tilde{\omega}_{i+1}}(\tilde{v},\tilde{\Omega}) = 0. \qquad (5.163)$$

The following is obtained:

$$\left[\frac{\partial J}{\partial \tilde{\omega}_i}(\tilde{v}, \tilde{\Omega}) = 0\right] \Leftrightarrow \cdots$$

$$\tilde{\Omega}_{i-1}\left[EI \int_{x_{i-1}}^{x_i} \dot{\varphi}_{i-1}\dot{\varphi}_i + \mu S \int_{x_{i-1}}^{x_i} \varphi_{i-1}\varphi_i\right] + \cdots$$

$$\tilde{\Omega}_i\left[EI \int_{x_{i-1}}^{x_i} \dot{\varphi}_i^2 + EI \int_{x_i}^{x_{i+1}} \dot{\varphi}_i^2\right.$$

$$\left. + \mu S \int_{x_{i-1}}^{x_i} \varphi_i^2 + \mu S \int_{x_i}^{x_{i+1}} \varphi_i^2\right] + \cdots$$

$$\tilde{\Omega}_{i+1}\left[EI \int_{x_i}^{x_{i+1}} \dot{\varphi}_i\dot{\varphi}_{i+1} + \mu S \int_{x_i}^{x_{i+1}} \varphi_i\varphi_{i+1}\right] + \cdots$$

$$\tilde{v}_{i-1}\left[-\mu S \int_{x_{i-1}}^{x_i} \dot{\varphi}_{i-1}\varphi_i\right] + \cdots$$

$$\tilde{v}_i\left[-\mu S \int_{x_{i-1}}^{x_i} \dot{\varphi}_i\varphi_i - \mu S \int_{x_i}^{x_{i+1}} \dot{\varphi}_i\varphi_i\right] + \cdots$$

$$\tilde{v}_{i+1}\left[-\mu S \int_{x_i}^{x_{i+1}} \dot{\varphi}_{i+1}\varphi_i\right] + \cdots$$

$$= 0 . \tag{5.164}$$

Then,

$$\left[\frac{\partial J}{\partial \tilde{\omega}_{i+1}}(\tilde{v}, \tilde{\Omega}) = 0\right] \Leftrightarrow \cdots$$

$$\tilde{\Omega}_i\left[EI \int_{x_i}^{x_{i+1}} \dot{\varphi}_i\dot{\varphi}_{i+1} + \mu S \int_{x_i}^{x_{i+1}} \varphi_i\varphi_{i+1}\right] + \cdots$$

$$\tilde{\Omega}_{i+1}\left[EI \int_{x_i}^{x_{i+1}} \dot{\varphi}_{i+1}^2 + EI \int_{x_{i+1}}^{x_{i+2}} \dot{\varphi}_{i+1}^2 + \mu S \int_{x_i}^{x_{i+1}} \varphi_{i+1}^2\right.$$

$$\left. + \mu S \int_{x_{i+1}}^{x_{i+2}} \varphi_{i+1}^2\right] + \cdots$$

$$\tilde{\Omega}_{i+2}\left[EI \int_{x_{i+1}}^{x_{i+2}} \dot{\varphi}_{i+1}\dot{\varphi}_{i+2} + \mu S \int_{x_{i+1}}^{x_{i+2}} \varphi_{i+1}\varphi_{i+2}\right] + \cdots$$

$$\tilde{v}_i\left[-\mu S \int_{x_i}^{x_{i+1}} \dot{\varphi}_i\varphi_{i+1}\right] + \cdots$$

$$\tilde{v}_{i+1}\left[-\mu S \int_{x_i}^{x_{i+1}} \varphi_{i+1}\dot{\varphi}_{i+1} - \mu S \int_{x_{i+1}}^{x_{i+2}} \varphi_{i+1}\dot{\varphi}_{i+1}\right] + \cdots$$

$$\tilde{v}_{i+2}\left[-\mu S \int_{x_{i+1}}^{x_{i+2}} \dot{\varphi}_{i+2}\varphi_{i+1}\right] + \cdots$$

$$= 0 . \tag{5.165}$$

It would have been noted that, for each group of equs. (5.161)–(5.162) and (5.164)–(5.165), the different contributions inherent to the finite element $[x_i, x_{i+1}]$ are in bold characters.

It is therefore sufficient to extract the weighting coefficients of the unknowns $(\tilde{v}_i, \tilde{v}_{i+1}, \tilde{\Omega}_i, \tilde{\Omega}_{i+1})$ from each of the four equations and to find their values by using the trapezium quadrature formula.

Therefore, the elementary matrix $A^{(i+1)}$ corresponds to:

$$
A^{(i+1)} = \begin{array}{cccc}
\tilde{v}_i \downarrow & \tilde{v}_{i+1} \downarrow & \tilde{\Omega}_i \downarrow & \tilde{\Omega}_{i+1} \downarrow \\
\end{array}
\left[\begin{array}{cccc}
\dfrac{\mu S}{h} & \dfrac{\mu S}{2} & -\dfrac{\mu S}{h} & \dfrac{\mu S}{2} \\[2mm]
\dfrac{\mu S}{2} & \dfrac{EI}{h} + \dfrac{\mu Sh}{2} & -\dfrac{\mu S}{2} & -\dfrac{EI}{h} \\[2mm]
-\dfrac{\mu S}{h} & -\dfrac{\mu S}{2} & \dfrac{\mu S}{h} & -\dfrac{\mu S}{2} \\[2mm]
\dfrac{\mu S}{2} & -\dfrac{EI}{h} & -\dfrac{\mu S}{2} & \dfrac{EI}{h} + \dfrac{\mu Sh}{2}
\end{array} \right]
\begin{array}{l}
\leftarrow \tilde{v}_i \\[2mm]
\leftarrow \tilde{v}_{i+1} \\[2mm]
\leftarrow \tilde{\Omega}_i \\[2mm]
\leftarrow \tilde{\Omega}_{i+1}
\end{array}
\tag{5.166}
$$

It can be noted that it is possible to rewrite matrix (5.166) in the form of (5.144), provided that the notations (5.115) are adopted.

Before initiating the assembly process and in order to distinguish the contribution of each mesh $[x_i, x_{i+1}]$ in the global matrix, the following writing norm is adopted by rewriting the local elementary matrix $A^{(i+1)}$ as:

$$
A^{(i+1)} = \begin{array}{cccc}
\tilde{v}_i \downarrow & \tilde{\Omega}_{i+1} \downarrow & \tilde{v}_{i+1} \downarrow & \tilde{\Omega}_{i+1} \downarrow \\
\end{array}
\left[\begin{array}{cccc}
\alpha^{(i+1)} & \beta^{(i+1)} & -\alpha^{(i+1)} & \beta^{(i+1)} \\[2mm]
\beta^{(i+1)} & \delta^{(i+1)} & -\beta^{(i+1)} & \gamma^{(i+1)} \\[2mm]
-\alpha^{(i+1)} & -\beta^{(i+1)} & \alpha^{(i+1)} & -\beta^{(i+1)} \\[2mm]
\beta^{(i+1)} & \gamma^{(i+1)} & -\beta^{(i+1)} & \delta^{(i+1)}
\end{array} \right]
\begin{array}{l}
\leftarrow \tilde{v}_i \\[2mm]
\leftarrow \tilde{\Omega}_i \\[2mm]
\leftarrow \tilde{v}_{i+1} \\[2mm]
\leftarrow \tilde{\Omega}_{i+1}
\end{array}
\tag{5.167}
$$

To avoid all confusion pertaining to the previous notations, it is pointed out that the notion of exponent in this present case only indicates, *in fine*, the contribution of each mesh $[x_i, x_{i+1}]$ in the global matrix.

However, as far as values of the coefficients matrix $A^{(i+1)}$ are concerned, they are constant and are not regulated by the finite element.

In other words:

$$
\forall i = 0, N: \ \alpha^{(i+1)} \equiv \alpha, \ \beta^{(i+1)} \equiv \beta, \ \gamma^{(i+1)} \equiv \gamma, \ \delta^{(i+1)} \equiv \delta, \tag{5.168}
$$

in which $(\alpha, \beta, \gamma, \delta)$ have been defined by (5.115).

A.13) This question is devoted to the process of assembly of the global matrix describing the linear system emanating from the minimisation problem $(\widetilde{\mathbf{MP}})$, whose unknowns are $(\tilde{v}_1, \tilde{\Omega}_1, \ldots, \tilde{v}_N, \tilde{\Omega}_N, \tilde{\Omega}_{N+1})$.

It is necessary to clearly note that for the node x_{N+1}, the approximate solution $\tilde{v}$ is zero at this point while the solution $\tilde{\Omega}$ possesses an unknown value $\tilde{\Omega}_{N+1}$ at the point x_{N+1}.

To constitute the global matrix of the linear system (5.107)–(5.108), each element $[x_i, x_{i+1}], (i = 0, N)$ of the mesh is passed and the global matrix is completed hop-by-hop with each of the local contributions.

The assembly process is initiated by considering the mesh $[x_0, x_1]$.

Concerning this first element, only the degrees of freedom at the abscissa x_1 must be considered since the beam is fitted when $x_0 = 0$.

In other words, if the local matrix $A^{(1)}$ is considered for this first element, only the sub-matrix relative to the unknowns $(\tilde{v}_1, \tilde{\Omega}_1)$ is to be considered.

This sub-matrix is presented in bold characters in the elementary matrix $A^{(1)}$, as shown below:

$$A^{(1)} = \begin{bmatrix} \alpha^{(1)} & \beta^{(1)} & -\alpha^{(1)} & \beta^{(1)} \\ \beta^{(1)} & \delta^{(1)} & -\beta^{(1)} & \gamma^{(1)} \\ -\boldsymbol{\alpha}^{(1)} & -\boldsymbol{\beta}^{(1)} & \boldsymbol{\alpha}^{(1)} & -\boldsymbol{\beta}^{(1)} \\ \boldsymbol{\beta}^{(1)} & \boldsymbol{\gamma}^{(1)} & -\boldsymbol{\beta}^{(1)} & \boldsymbol{\delta}^{(1)} \end{bmatrix} \begin{matrix} \\ \\ \leftarrow \tilde{v}_1 \\ \leftarrow \tilde{\Omega}_1 \end{matrix} \qquad (5.169)$$

with columns labelled $\tilde{v}_1 \downarrow \quad \tilde{\Omega}_1 \downarrow$.

The process is further performed with the second mesh $[x_1, x_2]$.

In this case, the elementary matrix $A^{(2)}$ is fully participates with its contribution, in so far as the two degrees of freedom, $(\tilde{v}_1, \tilde{\Omega}_1)$ on one hand and $(\tilde{v}_2, \tilde{\Omega}_2)$ on the other hand, are associated with the nodes x_1 and x_2.

Consequently, by maintaining the norm of bold characters for writing down the coefficients of matrix $A^{(2)}$ that would be considered during the assembly process, the matrix in question is written as:

$$A^{(2)} = \begin{bmatrix} \boldsymbol{\alpha}^{(2)} & \boldsymbol{\beta}^{(2)} & -\boldsymbol{\alpha}^{(2)} & \boldsymbol{\beta}^{(2)} \\ \boldsymbol{\beta}^{(2)} & \boldsymbol{\delta}^{(2)} & -\boldsymbol{\beta}^{(2)} & \boldsymbol{\gamma}^{(2)} \\ -\boldsymbol{\alpha}^{(2)} & -\boldsymbol{\beta}^{(2)} & \boldsymbol{\alpha}^{(2)} & -\boldsymbol{\beta}^{(2)} \\ \boldsymbol{\beta}^{(2)} & \boldsymbol{\gamma}^{(2)} & -\boldsymbol{\beta}^{(2)} & \boldsymbol{\delta}^{(2)} \end{bmatrix} \begin{matrix} \leftarrow \tilde{v}_1 \\ \leftarrow \tilde{\Omega}_1 \\ \leftarrow \tilde{v}_2 \\ \leftarrow \tilde{\Omega}_2 \end{matrix} \qquad (5.170)$$

with columns labelled $\tilde{v}_1 \downarrow \quad \tilde{\Omega}_1 \downarrow \quad \tilde{v}_2 \downarrow \quad \tilde{\Omega}_2 \downarrow$.

The contribution of the interval $[x_1,x_2]$ consequently extends to the other meshes $[x_i,x_{i+1}]$ for all values of i, ranging from 2 to N.

In other words, for the whole of the N meshes, the contribution of the elementary matrix $A^{(i+1)}$ is complete and by maintaining the same writing norms, the following is obtained:

$$
A^{(i+1)} =
\begin{array}{cccc}
\tilde{v}_i \quad & \tilde{\Omega}_i \quad & \tilde{v}_{i+1} \quad & \tilde{\Omega}_{i+1} \\
\downarrow & \downarrow & \downarrow & \downarrow
\end{array}
\begin{bmatrix}
\alpha^{(i+1)} & \beta^{(i+1)} & -\alpha^{(i+1)} & \beta^{(i+1)} \\
\beta^{(i+1)} & \delta^{(i+1)} & -\beta^{(i+1)} & \gamma^{(i+1)} \\
-\alpha^{(i+1)} & -\beta^{(i+1)} & \alpha^{(i+1)} & -\beta^{(i+1)} \\
\beta^{(i+1)} & \gamma^{(i+1)} & -\beta^{(i+1)} & \delta^{(i+1)}
\end{bmatrix}
\begin{array}{l}
\leftarrow \tilde{v}_i \\
\leftarrow \tilde{\Omega}_i \\
\leftarrow \tilde{v}_{i+1} \\
\leftarrow \tilde{\Omega}_{i+1}
\end{array}
\qquad (5.171)
$$

The integration of all of the elementary contributions of each finite element in the global matrix is then performed.

To easily visualise the global matrix structure, three levels of analysis are proposed: at the top left of the matrix, in its centre and at its bottom right.

Left upper corner of the stiffness matrix:

$$
A =
\begin{array}{cccccc}
\tilde{v}_1 & \tilde{\Omega}_1 & \tilde{v}_2 & \tilde{\Omega}_2 & \cdots & \tilde{\Omega}_{N+1} \\
\downarrow & \downarrow & \downarrow & \downarrow & & \downarrow
\end{array}
\begin{bmatrix}
\alpha^{(1)}+\alpha^{(2)} & -\beta^{(1)}+\beta^{(2)} & -\alpha^{(2)} & \beta^{(2)} & \cdots & \cdots \\
-\beta^{(1)}+\beta^{(2)} & \delta^{(1)}+\delta^{(2)} & -\beta^{(2)} & \gamma^{(2)} & \cdots & \cdots \\
-\alpha^{(2)} & -\beta^{(2)} & \alpha^{(2)}+\alpha^{(3)} & -\beta^{(2)}+\beta^{(3)} & \cdots & \cdots \\
\beta^{(2)} & \gamma^{(2)} & -\beta^{(2)}+\beta^{(3)} & \delta^{(2)}+\delta^{(3)} & \cdots & \cdots \\
\cdots & \cdots & \cdots & \cdots & \cdots & \cdots \\
\cdots & \cdots & \cdots & \cdots & \cdots & \cdots
\end{bmatrix}
\begin{array}{l}
\leftarrow \tilde{v}_1 \\
\leftarrow \tilde{\Omega}_1 \\
\leftarrow \tilde{v}_2 \\
\leftarrow \tilde{\Omega}_2 \\
\cdots \\
\leftarrow \tilde{\Omega}_{N+1}
\end{array}
$$

$$(5.172)$$

Generic centre of the stiffness matrix:

$$
\begin{array}{ccccc}
\cdots & \tilde{v}_i & \tilde{\Omega}_i & \tilde{v}_{i+1} & \tilde{\Omega}_{i+1} & \cdots \\
 & \downarrow & \downarrow & \downarrow & \downarrow &
\end{array}
$$

$$
A =
\begin{bmatrix}
\cdots & \cdots & \cdots & \cdots & \cdots & \cdots \\
\cdots & \alpha^{(i)}+\alpha^{(i+1)} & -\beta^{(i)}+\beta^{(i+1)} & -\alpha^{(i+1)} & \beta^{(i+1)} & \cdots \\
\cdots & -\beta^{(i)}+\beta^{(i+1)} & \delta^{(i)}+\delta^{(i+1)} & -\beta^{(i+1)} & \gamma^{(i+1)} & \cdots \\
 & -\alpha^{(i+1)} & -\beta^{(i+1)} & \alpha^{(i+1)}+\alpha^{(i+2)} & -\beta^{(i+1)}+\beta^{(i+2)} & \cdots \\
 & \beta^{(i+1)} & \gamma^{(i+1)} & -\beta^{(i+1)}+\beta^{(i+2)} & \delta^{(i+1)}+\delta^{(i+2)} & \cdots \\
\cdots & \cdots & \cdots & \cdots & \cdots & \cdots
\end{bmatrix}
\begin{array}{l}
\cdots \\
\leftarrow \tilde{v}_i \\
\leftarrow \tilde{\Omega}_i \\
\leftarrow \tilde{v}_{i+1} \\
\leftarrow \tilde{\Omega}_{i+1} \\
\cdots
\end{array}
$$

$$(5.173)$$

Lower right corner of the stiffness matrix:

$$
\begin{array}{cccc}
\cdots\ \cdots & \tilde{v}_N & \tilde{\Omega}_N & \tilde{\Omega}_{N+1} \\
 & \downarrow & \downarrow & \downarrow
\end{array}
$$

$$
A =
\begin{bmatrix}
\cdots\ \cdots & \cdots & \cdots & \cdots & \cdots \\
\cdots\ \cdots & \cdots & \cdots & \cdots & \cdots \\
\cdots\ \cdots & \alpha^{(N)}+\alpha^{(N+1)} & -\beta^{(N)}+\beta^{(N+1)} & \beta^{(N+1)} \\
\cdots\ \cdots & -\beta^{(N)}+\beta^{(N+1)} & \delta^{(N)}+\delta^{(N+1)} & \gamma^{(N+1)} \\
\cdots\ \cdots & \beta^{(N+1)} & \gamma^{(N+1)} & \delta^{(N+1)}
\end{bmatrix}
\begin{array}{l}
\cdots \\
\cdots \\
\leftarrow \tilde{v}_N \\
\leftarrow \tilde{\Omega}_N \\
\leftarrow \tilde{\Omega}_{N+1}
\end{array}
$$

$$(5.174)$$

The assembly is completed by performing all algebraic calculations and by considering the stationarity of the $\alpha^{(i)}$, $\beta^{(i)}$, $\gamma^{(i)}$ and $\delta^{(i)}$ sequences as indicated in (5.168).

The stiffness matrix then assumes the following final form:

$$
A =
\begin{array}{c}
\begin{array}{ccccccccccc}
\tilde{v}_1 & \tilde{\Omega}_1 & \tilde{v}_2 & \tilde{\Omega}_2 & \tilde{v}_3 & \tilde{\Omega}_3 & \tilde{v}_4 & \tilde{\Omega}_4 & \cdots & \tilde{v}_N & \tilde{\Omega}_N & \tilde{\Omega}_{N+1} \\
\downarrow & \downarrow & \downarrow & \downarrow & \downarrow & \downarrow & \downarrow & \downarrow & & \downarrow & \downarrow & \downarrow
\end{array} \\
\left[
\begin{array}{cccccccccccc}
2\alpha & 0 & -\alpha & \beta & & & & & \cdots & & & \\
0 & 2\delta & -\beta & \gamma & 0 & & & & \cdots & & & \\
-\alpha & -\beta & 2\alpha & 0 & -\alpha & \beta & & & \cdots & & & \\
\beta & \gamma & 0 & 2\delta & -\beta & \gamma & 0 & & \cdots & & & \\
& 0 & -\alpha & -\beta & 2\alpha & 0 & -\alpha & \beta & \cdots & & & \\
& & \beta & \gamma & 0 & 2\delta & -\beta & \gamma & \cdots & & & \\
& & & 0 & -\alpha & -\beta & 2\alpha & 0 & \cdots & & & \\
& & & & \beta & \gamma & 0 & 2\delta & \cdots & & & \\
\cdots & \cdots & \cdots & \cdots & \cdots & \cdots & \cdots & \cdots & & \cdots & \cdots & \cdots \\
& & & & & & & & \cdots & 2\alpha & 0 & \beta \\
& & & & & & & & \cdots & 0 & 2\delta & \gamma \\
& & & & & & & & \cdots & \beta & \gamma & \delta
\end{array}
\right]
\end{array}
\begin{array}{l}
\leftarrow \tilde{v}_1 \\
\leftarrow \tilde{\Omega}_1 \\
\leftarrow \tilde{v}_2 \\
\leftarrow \tilde{\Omega}_2 \\
\leftarrow \tilde{v}_3 \\
\leftarrow \tilde{\Omega}_3 \\
\leftarrow \tilde{v}_4 \\
\leftarrow \tilde{\Omega}_4 \\
\cdots \\
\leftarrow \tilde{v}_N \\
\leftarrow \tilde{\Omega}_N \\
\leftarrow \tilde{\Omega}_{N+1}
\end{array}
\qquad (5.175)
$$

5.2.2 Clamped-clamped beam – Euler-Bernoulli theory

Statement

Considering a homogenous beam Ω of length L, cross section S and constant density ρ whose mechanical behaviour is the isotropic elasticity with small disturbances. Moreover, it is noted that the constituent material of the beam has a Young's modulus of E and a Poisson's ratio of v.

The beam's axis is designated by $(O; \mathbf{X}_1)$ and the current abscissa is x. The beam's cross-section is parameterised by the geometric variables (y, z).

The beam is clamped-clamped at its ends when $x = 0$ and when $x = L$. Moreover, a volume density of transverse forces $\mathbf{f} = -f_2 \mathbf{X}_2$ is applied along the beam (see Fig. 5.7).

The density f_2 is given and has all the functional regularity properties so as to perform the integration calculations of the first two questions of the theoretical part.

▶ **Displacements Variational Formulation – Principle of Virtual Works – Theoretical Part**

1) The kinematic frame chosen to describe the real displacement field is that of Euler-Bernoulli. This justifies choosing to consider virtual displacement fields $\mathbf{U}^*$ expressed in the form:

$$\mathbf{U}^* = -y\dot{u}^*(x)\mathbf{X}_1 + u^*(x)\mathbf{X}_2 , \qquad (5.176)$$

satisfying boundary conditions:

$$u^*(0) = u^*(L) = 0 , \qquad (5.177)$$
$$\dot{u}^*(0) = \dot{u}^*(L) = 0 , \qquad (5.178)$$

where the following expression has been adopted: $\dot{u}^* \equiv \dfrac{du^*}{dx}$.

In addition, (5.177) and (5.178) demonstrate the clamping condition for any virtual field $\mathbf{U}^*$.

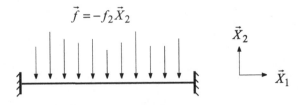

Fig. 5.7 Clamped-clamped Beam Subject to Flexion Forces

$\mathfrak{U}^*$ denotes the totality of the displacement fields satisfying all three (5.176) to (5.178) conditions.

The virtual work of the internal forces T_{Int}^*, and the virtual work of external forces T_{Ext}^* are determined and defined by:

$$T_{\text{Int}}^* = -\int_{\Omega} \sigma_{ij}\varepsilon_{ij}(\mathbf{U}^*)\,d\Omega, \quad T_{\text{Ext}}^* = \int_{\Omega} \mathbf{f}\cdot\mathbf{U}^*\,d\Omega, \tag{5.179}$$

where σ denotes the stress tensor and $\varepsilon(\mathbf{U}^*)$ the linear strain tensor associated with virtual displacement fields $\mathbf{U}^*$:

$$\varepsilon_{ij}(\mathbf{U}^*) = \frac{1}{2}\left[\frac{\partial U_i^*}{\partial x_j} + \frac{\partial U_j^*}{\partial x_i}\right]. \tag{5.180}$$

The bending moment M is to be introduced together with the 1D-density of loading forces f and is defined by:

$$M(x) = -\iint_{S(x)} y\sigma_{11}\,dS(x), \quad f(x) = \iint_{S(x)} f_2\,dS(x), \tag{5.181}$$

where $S(x)$ denotes abscissa x.

2) Without justifying the convergence of the concerned integrals, show that applying the principle of virtual works leads to the following formal variational formulation **(EVP)**:

$$(\textbf{EVP}) \quad \begin{bmatrix} \text{Find } M \text{ defined from } [0,L] \text{ to } \mathbf{R}, \text{ solution of:} \\ \int_0^L \left[\ddot{M}(x)+f\right]u^*(x)\,dx = 0, \forall u^* \in \mathfrak{U}^*. \end{bmatrix} \tag{5.182}$$

3) Using the behaviour law of the beam's constituent material $(M(x) = EI\ddot{u}(x)$, I being the beam's moment of inertia in relation to the $(O;\mathbf{X}_3)$ axis), and assuming that the force density f belongs to $L^2(0,L)$, show that the continuous problem **(CP)** composed of the beam's equilibrium equations is expressed as:

$$(\textbf{CP}) \quad \begin{bmatrix} \text{Find } u \text{ belonging to } H^4([0,L]), \text{ solution of:} \\ -EIu^{(4)}(x) = f(x), \quad \forall x \in]0,L[, \\ u(0) = u(L) = 0, \\ \dot{u}(0) = \dot{u}(L) = 0, \end{bmatrix} \tag{5.183}$$

where the Sobolev space $H^4([0,L])$ is defined by:

$$H^4(0,L) = \{v: [0,L] \to \mathbf{R}, v^{(k)} \in L^2(0,L), (k = 0 \text{ to } 4)\}. \tag{5.184}$$

4) $V \equiv H_0^2([0,L])$ denotes the Sobolev space defined by:

$$H_0^2(0,L) = \{v/v^{(k)} \in L^2(0,L), \ (k = 0 \text{ to } 2)\} \cap W_0 \,,$$

where:

$$W_0 = \{v/v(0) = \dot{v}(0) = v(L) = \dot{v}(L) = 0\} \,. \tag{5.185}$$

Then, show that a displacements variational formulation (**VP**) can be expressed as:

$$(\textbf{VP}) \quad \begin{bmatrix} \text{Find } u \text{ belonging to } V, \text{ solution of:} \\[2mm] EI \int_0^L \ddot{u}(x)\ddot{v}(x)\,dx + \int_0^L f(x)v(x)\,dx = 0 \,, \quad \forall v \in V \,. \end{bmatrix} \tag{5.186}$$

5) Show that there is a minimization problem (**MP**) equivalent to the variational formulation (**VP**) defined by:

$$(\textbf{MP}) \quad \begin{bmatrix} \text{Find } u \text{ belonging to } V, \text{ solution of:} \quad J(u) = \underset{v \in V}{\text{Min }} J(v) \,, \\[3mm] \text{where:} \quad J(v) = \dfrac{EI}{2} \int_0^L \ddot{v}^2(x)\,dx + \int_0^L f(x)v(x)\,dx \,. \end{bmatrix} \tag{5.187}$$

▶ **Numerical Part**

This part is dedicated to the approximation by applied finite elements of the elastic beams in flexion deformation, modelised by the Euler-Bernoulli theory.

To achieve this, the Hermite finite elements can be applied as follows:

Let the regular mesh be at interval $[0,L]$ and of constant step h, such as:

$$\begin{cases} x_0 = 0, \ x_{N+1} = L \,, \\ x_{i+1} = x_i + h, \quad i = 0 \text{ to } N \,. \end{cases} \tag{5.188}$$

In addition, let the approximation space $\tilde{W}$ be defined by:

$$\tilde{W} = \{\tilde{w}: [0,L] \rightarrow \mathbf{R}, \quad \tilde{w} \in C^1([0,L]), \quad \tilde{w}|_{[x_i,x_{i+1}]} \in P_3([x_i,x_{i+1}])\} \,, \tag{5.189}$$

where $P_3([x_i, xi_{+1}])$ denotes the polynomial space defined on $[x_i, x_{i+1}]$, of degree less than or equal to three.

In the same manner, introduce space $\tilde{V}$ defined by:

$$\tilde{V} = \{\tilde{v} \in \tilde{W} / \tilde{v}(0) = \dot{\tilde{v}}(0) = \tilde{v}(L) = \dot{\tilde{v}}(L) = 0\} \,. \tag{5.190}$$

6) What are the dimensions of the $\tilde{W}$ and $\tilde{V}$ spaces?

7) The system of functions $(\varphi_i)_{i=1,2N}$ of $\tilde{V}$, divided in two groups, is considered as follows:

$$\dot{\varphi}_{2k}(x_j) = \delta_{2k,j}, \varphi_{2k}(x_j) = 0, \quad (\forall j = 1 \text{ to } N) \tag{5.191}$$

$$\varphi_{2k+1}(x_j) = \delta_{2k+1,j}, \dot{\varphi}_{2k+1}(x_j) = 0, \quad (\forall j = 1 \text{ to } N). \tag{5.192}$$

- What is the support of functions $(\varphi_i)_{i=1,2N}$?
- Show that the $2N$ functions $(\varphi_i)_{i=1,2N}$ constitute a basis of the space $\tilde{V}$.

8) Let $\tilde{v}$ belonging to $\tilde{V}$ be defined by:

$$\tilde{v} = \sum_{k=1}^{N} \tilde{\alpha}_k \varphi_{2k} + \sum_{k=1}^{N} \tilde{\beta}_k \varphi_{2k-1}. \tag{5.193}$$

- Suggest a mechanical interpretation of the $\tilde{\alpha}_k$ and $\tilde{\beta}_k$ coefficients.
- Explain how the approximation $\tilde{v}$ of a mesh $[x_i, x_{i+1}]$ is expressed?
- State the 4 basis functions φ_{2i-1}, φ_{2i}, φ_{2i+1} and φ_{2i+2} on element $[x_i x_{i+1}]$.

9) Let $(\widetilde{MP})$ be the approximate minimization problem, associated to the (MP) problem, defined by:

$(\widetilde{MP})$ Find $\tilde{u}$ belonging to $\tilde{V}$, being solution to: $\quad J(\tilde{u}) = \underset{\tilde{v} \in \tilde{V}}{\text{Min}} J(\tilde{v})$,

where J is the potential energy defined by (5.187).

- Show that a necessary condition for $\tilde{u}$ to be a solution to $(\widetilde{MP})$ is written as:

Determine the $(\tilde{u}_i)_{1,N}$ and $(\dot{\tilde{u}}_i)_{1,N}$ numerical sequences, defining approximation $\tilde{u}$ belonging to $\tilde{V}$ solution to the global linear system:

$$
(\widetilde{MP}) \quad
\begin{cases}
\forall i = 1 \text{ to } N: \\[2mm]
EI \sum_{k=1}^{N} \left[\int_{S_{2i-1,2k}} \ddot{\varphi}_{2i-1} \ddot{\varphi}_{2k} \, dx \right] \dot{\tilde{u}}_k + \sum_{k=1}^{N} \left[\int_{S_{2i-1,2k-1}} \ddot{\varphi}_{2i-1} \ddot{\varphi}_{2k-1} \, dx \right] \tilde{u}_k \\[4mm]
\qquad = -\int_{S_{2i-1}} f \varphi_{2i-1} \, dx, \tag{5.194} \\[4mm]
EI \sum_{k=1}^{N} \left[\int_{S_{2i,2k}} \ddot{\varphi}_{2i} \ddot{\varphi}_{2k} \, dx \right] \dot{\tilde{u}}_k + \sum_{k=1}^{N} \left[\int_{S_{2i,2k-1}} \ddot{\varphi}_{2i} \ddot{\varphi}_{2k-1} \, dx \right] \tilde{u}_k \\[4mm]
\qquad = -\int_{S_{2i}} f \varphi_{2i} \, dx, \tag{5.195}
\end{cases}
$$

where the following notations have been adopted:

$$S_{l,m} \equiv \text{Supp } \varphi_l \cap \text{Supp } \varphi_m, \quad S_m \equiv \text{Supp } \varphi_m. \tag{5.196}$$

10) The elementary matrix $a^{(i+1)}$ is now introduced, showing the contribution of segment $[x_i, x_{i+1}]$ in the global matrix A of the linear system (5.194) to (5.195), namely the $(i+1)^{\text{th}}$ mesh element from interval $[0, L]$.

Thus, the result obtained is:

$$a^{(i+1)} = \frac{2EI}{h^3} \begin{bmatrix} 6 & 3h & -6 & 3h \\ 3h & 2h^2 & -3h & h^2 \\ -6 & -3h & 6 & -3h \\ 3h & h^2 & -3h & 2h^2 \end{bmatrix}. \tag{5.197}$$

The elementary vector $b^{(i+1)}$ is introduced in the same way and is defined by:

$$b^{(i+1)} = \begin{bmatrix} b_1^{(i+1)} \\ b_2^{(i+1)} \\ b_3^{(i+1)} \\ b_4^{(i+1)} \end{bmatrix} \equiv \begin{bmatrix} -\int_{x_i}^{x_{i+1}} f\varphi_{2i-1}\, dx \\ -\int_{x_i}^{x_{i+1}} f\varphi_{2i}\, dx \\ -\int_{x_i}^{x_{i+1}} f\varphi_{2i+1}\, dx \\ -\int_{x_i}^{x_{i+1}} f\varphi_{2i+2}\, dx \end{bmatrix}. \tag{5.198}$$

– Using the Simpson quadrature formula, suggest an approximation for vector $b^{(i+1)}$.

For reminder, the Simpson formula is expressed as:

$$\int_a^b f(x)\, dx \simeq \frac{(b-a)}{6}\left[f(a) + 4f\left(\frac{a+b}{2}\right) + f(b) \right]. \tag{5.199}$$

– Show that, in the special case of a uniformly distributed load p, $(\mathbf{f} = -f_2\mathbf{X}_2 \equiv -p\mathbf{X}_2)$, the result for vector $b_*^{(i+1)}$ is found again and is defined by:

$${}^t b_*^{(i+1)} = \left[-\frac{ph}{2}, -\frac{ph^2}{12}, \frac{ph}{2}, \frac{ph^2}{12} \right]. \tag{5.200}$$

11) Consider a special case where the beam's mesh Ω is composed of 3 meshes of constant length h.

– Process the assembly of the global matrix A with generic element A_{ij} as well as that of the second member b of component b_i corresponding to problem $(\widetilde{\mathbf{MP}})$.

– Infer the form of the approximation of the displacement field at all points of beam Ω.

– Are the nodal equations obtained by the classical approximation techniques using Hermite finite elements applied to the variational formulation (**VP**), seen again?

_____

5.2.3 Solution

▶ **Displacements Variational Formulation – Principle of Virtual Works – Theoretical Part**

A.1) The structure of a virtual displacements field $\mathbf{U}^*$ defined by (5.176) implies that the strain tensor $\varepsilon(\mathbf{U}^*)$ (cf. (5.180)) associated with $\mathbf{U}^*$ is written as:

$$\varepsilon(\mathbf{U}^*) = \varepsilon_{11}^* \mathbf{X}_1 \otimes \mathbf{X}_1 = -y\ddot{u}^*(x)\mathbf{X}_1 \otimes \mathbf{X}_1 . \qquad (5.201)$$

The notation of tensor product $\mathbf{X}_1 \otimes \mathbf{X}_1$ expressing the fact that the tensor $\varepsilon(\mathbf{U}^*)$ has only one non-zero component, namely ε_{11}^* would have been introduced.

The virtual work of internal forces T_{Int}^* is evaluated in the following way:

$$T_{\text{Int}}^* = -\int_{\Omega} \sigma_{ij}\varepsilon_{ij}(\mathbf{U}^*)\,d\Omega = -\int_{\Omega} \sigma_{11}\varepsilon_{11}(\mathbf{U}^*)\,d\Omega ,$$

$$= -\int_0^L \left(-\int\int_{S(x)} y\sigma_{11}\,dS(x)\right)\ddot{u}^*(x)\,dx = -\int_0^L M(x)\ddot{u}^*(x)\,dx, \qquad (5.202)$$

where the definition of the bending moment M defined in (5.181) has been used.

In an analogous way, the virtual work of external forces is obtained from:

$$T_{\text{Ext}}^* = \int_{\Omega} \mathbf{f}\cdot\mathbf{U}^*\,d\Omega = -\int_{\Omega} f_2 u^*(x)\,d\Omega$$

$$= -\int_0^L \left(\int\int_{S(x)} f_2\,dS(x)\right)u^*(x)\,dx = -\int_0^L f(x)u^*(x)\,dx . \qquad (5.203)$$

A.2) The application of the principle of virtual works in the case of an elastostatic problem is written as:

$$T_{\text{Int}}^* + T_{\text{Ext}}^* = 0 , \quad \forall u^* \in \mathfrak{U}^* . \qquad (5.204)$$

Therefore, by using the expressions (5.202) and (5.203), the following is obtained:

$$\int_0^L M(x)\ddot{u}^*(x)\,dx + \int_0^L f(x)u^*(x)\,dx = 0 , \quad \forall u^* \in \mathfrak{U}^* . \qquad (5.205)$$

A double integration by parts transforms the first integral of (5.205) and gives:

$$\int_0^L [\ddot{M}(x) + f(x)]\,u^*(x)\,dx - [\dot{M}u^*]_0^L + [M\dot{u}^*]_0^L = 0 , \quad \forall u^* \in \mathfrak{U}^* . \qquad (5.206)$$

Consequently, by using the boundary conditions (5.177) and (5.178) of the virtual fields of $\mathfrak{U}^*$, the formal variational formulation (**EVP**) is obtained:

$$(\textbf{EVP}) \quad \begin{bmatrix} \text{Find } M \text{ defined from } [0,L] \text{ to } \mathbf{R} \text{ being the solution of:} \\ \int_0^L [\ddot{M}(x) + f(x)]\,u^*(x)\,dx = 0 , \quad \forall u^* \in \mathfrak{U}^* . \end{bmatrix} \qquad (5.207)$$

A.3) Solving of the continuous problem (**CP**) is worked out from the density arguments as applied on several previous occasions.

To achieve this, formulation (5.207) of the variational equation (**EVP**) is applied again and the bending moment M is substituted by its expression according to the displacement field of solution u via the behaviour law mentioned in the statement of this question.

The following is then obtained:

$$\int_0^L \left[EIu^{(4)}(x) + f(x)\right] u^*(x)\,dx = 0\,, \quad \forall u^* \in \mathfrak{U}^*\,. \tag{5.208}$$

To give a sense to the integration of variational formulation (5.208), the solution u would be searched within the space $H^4(0,L)$ defined by (5.184).

Indeed, considering that the fact f belongs to $L^2(0,L)$, should the fourth derivative $u^{(4)}$ as well as the virtual field u^* be set to belong to $L^2(0,L)$, the existence of the integral of (5.208) would as such be ensured.

The stress pertaining to the boundary conditions bearing upon u and u^* are added to these regularity conditions.

It can then be considered that the solution u and the virtual fields u^* belong to the Sobolev space $H^4_*([0;L])$ defined by:

$$H^4_*(0,L) \equiv H^4(0,L) \cap \mathfrak{U}^*\,. \tag{5.209}$$

The variational formulation (**EVP**) is consequently written as:

$$(\textbf{EVP}) \quad \left[\begin{array}{l} \text{Find } u \text{ belonging to } H^4_*(0,L) \text{ which is the solution of:} \\[2mm] \displaystyle\int_0^L \left[EIu^{(4)}(x) + f(x)\right] u^*(x)\,dx = 0\,, \quad \forall u^* \in H^4_*(0,L)\,. \end{array}\right. \tag{5.210}$$

The continuous problem (**CP**) is then estimated by using the density method:

Among all the functions of $H^4_*(0,L)$, those which belong to $\mathscr{D}(0,L)$ are considered (since $\mathscr{D}(0,L) \subset H^4_*(0,L)$) and the fact that $\mathscr{D}(0,L)$ is dense in $L^2(0,L)$ is used.

The reader may refer to the elaboration of the density method as applied in the case of the problem of a beam subjected to traction [5.1.1].

The following is then easily obtained:

$$EIu^{(4)}(x) + f(x) = 0\,, \quad \forall x \in]0,L[\,. \tag{5.211}$$

▶ **Summary**

The solution u of the variational equation (5.210) has been considered in the space $H^4_*(0,L)$.

By using the density method, it has then been proven that u satisfies differential equation (5.211), while satisfying the homogenous boundary conditions of Dirichlet and Neumann when $x = 0$ and when $x = L$, which are the properties at the border of the interval $[0, L]$ of the functions of $H_*^4(0, L)$.

In other words, u is the solution of the continuous problem (**CP**) defined by (5.183).

Finally, it can be noted that when the distribution of forces f shows more regularity, at least continuous along the interval $[0, L]$, the solution u of the continuous problem (**CP**) is then the classical solution belonging to $C^4(]0, L[)$.

A.4) To prove the variational formulation (**VP**), let v be a function belonging to a variational space V to be later defined.

The fourth orde r differential equation (5.183) is multiplied, and two successive integrations by parts is performed.

The following formulation is then found:

$$EI \int_0^L [\ddot{u}\ddot{v} + fv]\, dx + \left[u^{(3)} v \right]_0^L - [\ddot{u}\dot{v}]_0^L = 0, \quad \forall v \in V. \qquad (5.212)$$

The definition of the functional framework V occurring in formulation (5.212) can now be studied.

Firstly, it can be observed that subsequent to the double integration by parts, variational equation (5.212) does not explicitly manifest the values of solution u and of its derivative $\dot{u}$ when $x = 0$ and when $x = L$.

In order to record this information in the variational formulation (**VP**), it is compulsory that the functions v of V (area of investigation of solution u) satisfy the following boundary conditions:

$$v(0) = \dot{v}(0) = v(L) = \dot{v}(L) = 0. \qquad (5.213)$$

Thus, the future solution u also belonging to V as one of the functions v of V, will satisfy *de facto* the boundary conditions of the clamped-clamped beam Ω.

Consequently, equation (5.212) is then written as:

$$EI \int_0^L [\ddot{u}\ddot{v} + fv]\, dx = 0, \quad \forall v \in V. \qquad (5.214)$$

With regards to the regularity of the functions v of V, the Sobolev space $H^2(0, L)$ will be considered so as to ensure the convergence of the two integrals of equation (5.214).

The variational formulation (**VP**) as defined in (5.186) is hence obtained.

A.5) The variational formulation (**VP**) defined by (5.186) comprises a bilinear form $a(.,.)$ and a linear form $L(.)$ respectively defined by:

$$a(u,v) = EI \int_0^L \ddot{u}\ddot{v}\,dx\,, \quad L(v) = -\int_0^L fv\,dx\,, \tag{5.215}$$

in which u and v describe $H_0^2(0,L)$.

Thus, the variational formulation can be written in a generic form:

Find $u \in V$ being a solution of:

$$a(u,v) = L(v)\,, \quad \forall v \in V\,. \tag{5.216}$$

Furthermore, it is clear that the bilinear form a is symmetrical and positive. These properties ensure that an equivalent minimisation problem defined by (**MP**) exists, (cf. Raviart [7] or Euvrard [4]).

▶ **Numerical Part**

The dimension of space $\tilde{W}$ is equal to $2N+4$ while that of space $\tilde{V}$ is equal to $2N$.

Indeed, to determine the dimension of $\tilde{W}$, it suffices to calculate the number of degrees of freedom that characterise any function $\tilde{w}$, belonging to $\tilde{W}$.

However, any function $\tilde{w}$ of $\tilde{W}$ is a third degree polynomial by mesh $[x_i, x_{i+1}]$. Therefore, by using mesh $[x_i, x_{i+1}]$, four degrees of freedom are available. In such a way that over the whole of the $(N+1)$ meshes, $4(N+1)$ degrees of freedom are obtained.

There still is need to eliminate the continuity conditions resulting from the junction of two adjacent meshes for both function $\tilde{\omega}$ as well as $\dot{\tilde{\omega}}$ for its derivative in so far as the functions of $\tilde{W}$ are C^1 along the interval $[0,L]$.

In other words, the $2N$ continuity conditions for $\tilde{w}$ and $\dot{\tilde{w}}$ correspond to the N points of the junction $(x_1, \ldots, x_N)$.

Finally, any function $\tilde{w}$ has $4(N+1) - 2N$ degrees of freedom, that is $(2N+4)$ in all and $\dim \tilde{W} = 2N+4$.

The dimension of $\tilde{V}$ can be immediately inferred since the fixed values at zero of $\tilde{v}$ and of $\dot{\tilde{v}}$ when $x=0$ and when $x=L$ need to be considered, therefore scaling any function $\tilde{v}$ of $\tilde{V}$ by four degrees of freedom.

This explains why the dimension of $\tilde{V}$ is equal to $2N$.

A.7) Let φ_i be one of the $2N$ functions of $\tilde{V}$ defined by the conditions (5.191)–(5.192).

a) *Study of the support of functions* φ_i.

For any interval $[x_k, x_{k+1}]$, which is different from $[x_{i-1}, x_i]$ or from $[x_i, x_{i+1}]$, φ_i is a function that is zero like its derivative at points x_k and x_{k+1}.

Indeed, given that function φ_i is a polynomial having a degree less or equal to three along the interval $[x_k, x_{k+1}]$, this polynomial is bound to have a zero value along the whole of the interval.

To prove this, it is only necessary to assume, to the point of absurdity, that such a function has a non zero value along the interval $[x_k, x_{k+1}]$.

In this case, given the boundary conditions when using x_k and x_{k+1}, this function would necessary take one of the following two forms (cf. Fig. 5.8):

It is then seen that the profile of φ_i inevitably shows two or three inflexion points; such an aspect does not occur in a third degree polynomial.

A more analytical demonstration consists in applying the Rolle's theorem, given the fact that:

$$\varphi_i(x_k) = \varphi_i(x_{k+1}) = 0 \quad \text{and} \quad \dot{\varphi}_i(x_k) = \dot{\varphi}_i(x_{k+1}) = 0 .$$

The consequent result is that the support of functions φ_i is the union of intervals $[x_{i-1}, x_i]$ and $[x_i, x_{i+1}]$.

b) *Study of the system of the 2N functions* $(\varphi_i)_{i=1,2N}$ *of* $\tilde{V}$.

To prove that the 2N functions $(\varphi_i)_{i=1,2N}$ constitute a basis of $\tilde{V}$, it is suggested to prove that it constitutes a free system of 2N elements in a vector space having a dimension of 2N.

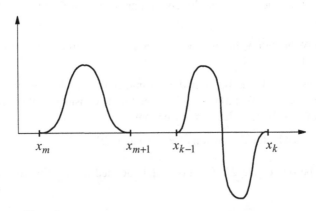

Fig. 5.8 Counter-Example

To achieve this, two sequences of real coefficients $(\xi_i)_{i=1,N}$ and $(\eta_i)_{i=1,N}$ are chosen and the linear combination is considered:

$$\sum_{k=1}^{N} \xi_k \varphi_{2k} + \sum_{k=1}^{N} \eta_k \varphi_{2k-1} = 0 \, . \tag{5.217}$$

By evaluating (5.217) at nodes x_i, it is seen that all coefficients η_i have a zero value. Furthermore, following the derivation of (5.217), at point x_i once more, it is similarly observed that the coefficients ξ_i also have a zero value.

It would be observed that such a result is essentially based on properties (5.191)–(5.192) of functions φ_i and of their derivative $\dot{\varphi}_i$.

The family of $2N$ functions φ_i is linearly independent; it is therefore a generator in a vector space of dimension $2N$ and constitutes, as a result, a basis of $\tilde{V}$.

A.8) The approximation proposed in this problem pertains to the Hermite's finite elements.

The aim is to carry out an interpolation, of the displacement field $\tilde{u}$ and of its derivative $\dot{\tilde{u}}$, which represents the rotation of the plane section according to the Euler-Bernoulli theory whereas in the Timoshenko beam theory, rotation θ of the section of the beam is an independent function of the displacement field $\mathbf{U}$.

a) *Mechanical interpretation of coefficients $\tilde{\alpha}_k$ and $\tilde{\beta}_k$.*

Any function $\tilde{v}$ of $\tilde{V}$ breaks down over the two families of basis functions $(\varphi_{2k})_{k=1,N}$ and $(\varphi_{2k-1})_{k=1,N}$ in such a way that:

$$\tilde{v} = \sum_{k=1}^{N} \tilde{\alpha}_k \varphi_{2k} + \sum_{k=1}^{N} \tilde{\beta}_k \varphi_{2k-1} \, . \tag{5.218}$$

Therefore, if factorisation (5.218) is evaluated at point x_i, considering the properties of the basis functions (5.191)–(5.192), the following is obtained:

$$\tilde{v}(x_i) = \tilde{\beta}_i \, . \tag{5.219}$$

Similarly, after having derived (5.217), the derivative of $\tilde{v}$ is calculated at point x_i and the following is obtained:

$$\dot{\tilde{v}}(x_i) = \tilde{\alpha}_i \, . \tag{5.220}$$

This explains why coefficients α_i and β_i are respectively interpreted as the approximation of derivative $\dot{\tilde{v}}_i$, (the rotation of the section) and of the displacement $\tilde{v}_i$ at node x_i.

The approximation $\tilde{v}$ is finally expressed as:

$$\tilde{v} = \sum_{k=1}^{N} \dot{\tilde{v}}_k \varphi_{2k} + \sum_{k=1}^{N} \tilde{v}_k \varphi_{2k-1} \, , \tag{5.221}$$

where it has been set that:

$$\tilde{v}_k \equiv \tilde{v}(x_k) \quad \text{and} \quad \dot{\tilde{v}}_k \equiv \dot{\tilde{v}}(x_k) . \tag{5.222}$$

b) *Restriction of the approximation $\tilde{v}$ to a mesh $[x_i, x_{i+1}]$.*

In the previous question, it has been proven that the support of functions $(\varphi_i)_{i=1,2N}$ is constituted by the union of intervals $[x_{i-1}, x_i]$ and $[x_i, x_{i+1}]$.

Therefore, on mesh $[x_i, x_{i+1}]$, the restriction of a function $\tilde{v}$ of $\tilde{V}$ is expressed as:

$$\tilde{v} = \tilde{v}_i \varphi_{2i-1} + \dot{\tilde{v}}_i \varphi_{2i} + \tilde{v}_{i+1} \varphi_{2i+1} + \dot{\tilde{v}}_{i+1} \varphi_{2i+2} . \tag{5.223}$$

c) *Expression of basis functions on the mesh $[x_i, x_{i+1}]$.*

The only basis functions having part of their support along the interval $[x_i, x_{i+1}]$ are:

$$\varphi_{2i-1} , \quad \varphi_{2i} , \quad \varphi_{2i+1} \quad \text{and} \quad \varphi_{2i+2} .$$

These four functions are third degree polynomials and must satisfy properties (5.191)–(5.192).

Therefore, function $\varphi_{2i-1}(x)$ is defined by:

$$\varphi_{2i-1}(x_i) = 1 , \quad \dot{\varphi}_{2i-1}(x_i) = 0 , \tag{5.224}$$

$$\varphi_{2i-1}(x_{i+1}) = 0 , \quad \dot{\varphi}_{2i-1}(x_{i+1}) = 0 . \tag{5.225}$$

Considering the cubic polynomial structure of φ_{2i-1}, conditions (5.225) imply that is expressed as:

$$\varphi_{2i-1}(x) = (x - x_{i+1})^2 (Ax + B) . \tag{5.226}$$

Furthermore, the boundary conditions (5.224) enable the evaluation of coefficients A and B and the function φ_{2i-1} is expressed in its final form as:

$$\varphi_{2i-1}(x) = \frac{1}{h^3} (x - x_{i+1})^2 [2(x - x_i) + h] . \tag{5.227}$$

Similarly, an analogous reasoning leads to the expression of the other basis functions:

$$\varphi_{2i}(x) = \frac{1}{h^2} (x - x_{i+1})^2 (x - x_i) , \tag{5.228}$$

$$\varphi_{2i+1}(x) = \frac{1}{h^3} (x - x_i)^2 [2(x_{i+1} - x) + h] , \tag{5.229}$$

$$\varphi_{2i+2}(x) = \frac{1}{h^2} (x - x_i)^2 (x - x_{i+1}) . \tag{5.230}$$

A.9) To obtain the approximate formulation of the minimisation problem (**MP**), the functional J defined in (5.187) is evaluated at point $\tilde{v}$ obtained from the expansion (5.193).

To simplify the written expressions, the mute index summation (or Einstein summation convention), is adopted and the following is obtained:

$$J(\tilde{v}) = \frac{EI}{2} \int_0^L \left[\dot{\tilde{v}}_k \ddot{\varphi}_{2k} + \tilde{v}_k \ddot{\varphi}_{2k-1} \right]^2 dx + \int_0^L f \left[\dot{\tilde{v}}_k \varphi_{2k} + \tilde{v}_k \varphi_{2k-1} \right] dx. \qquad (5.231)$$

A necessary condition for minimisation of the functional J is therefore written as:

$$\forall k = 1 \text{ to } N : \frac{\partial J}{\partial \dot{\tilde{v}}_k}(\tilde{u}) = \frac{\partial J}{\partial \tilde{v}_k}(\tilde{u}) = 0. \qquad (5.232)$$

The $2N$ conditions (5.232) are therefore expressed as:

$$(\widetilde{MP}) \quad \left[\begin{array}{l} EI \int_0^L \left[\ddot{\tilde{u}}_k \ddot{\varphi}_{2k} + \tilde{u}_k \ddot{\varphi}_{2k-1} \right] \ddot{\varphi}_{2k-1} \, dx = - \int_0^L f \varphi_{2k-1} \, dx, \\[4pt] (\forall k = 1 \text{ to } N), \\[8pt] EI \int_0^L \left[\ddot{\tilde{u}}_k \ddot{\varphi}_{2k} + \tilde{u}_{k+1} \ddot{\varphi}_{2k+1} \right] \ddot{\varphi}_{2k} \, dx = - \int_0^L f \varphi_{2k} \, dx, \\[4pt] (\forall k = 1 \text{ to } N). \end{array} \right. \qquad \begin{array}{l} (5.233) \\[40pt] (5.234) \end{array}$$

It is finally observed that the integrands occurring in the two equations (5.233) and (5.234) exclusively bear upon the intersection of the supports of the basis functions $(\varphi_i)_{i=1,2N}$, resulting in formulation (5.194)–(5.195).

A.10) The elementary matrix $a^{(i+1)}$ defined by (5.197) integrates the contribution of element $[x_i, x_{i+1}]$ in the global matrix A of system (5.194)–(5.195).

Similarly, $b^{(i+1)}$ is the contribution of this same element in the second member b.

a) *Approximation of the Elementary Vector $b^{(i+1)}$ by the Simpson formula.*

Each of the 4 components of vector $b^{(i+1)}$ is evaluated by approximation using the Simpson quadrature formula (5.199).

$$\begin{aligned} b_1^{(i+1)} &= - \int_{x_i}^{x_{i+1}} f \varphi_{2i-1} \, dx = - \int_{x_i}^{x_{i+1}} \frac{1}{h^3} (x - x_{i+1})^2 [2(x - x_i) + h] f(x) \, dx \\ &\simeq - \frac{h}{6h^3} \left(f_i h^3 + 4(x_{i+\frac{1}{2}} - x_{i+1})^2 \left[2 \left(x_{i+\frac{1}{2}} - x_i \right) + h \right] f_{i+\frac{1}{2}} \right) \\ &\simeq - \frac{h}{6} \left(f_i + 2 f_{i+\frac{1}{2}} \right). \end{aligned} \qquad (5.235)$$

Likewise, the following is obtained:

$$\begin{aligned} b_2^{(i+1)} &= - \int_{x_i}^{x_{i+1}} f \varphi_{2i} \, dx = - \int_{x_i}^{x_{i+1}} \frac{1}{h^2} (x - x_{i+1})^2 (x - x_i) f(x) \, dx \\ &\simeq - \frac{h}{6h^2} \left((0 \times f_i) + 4 \left(x_{i+\frac{1}{2}} - x_{i+1} \right)^2 \left(x_{i+\frac{1}{2}} - x_i \right) f_{i+\frac{1}{2}} + (0 \times f_{i+1}) \right) \\ &\simeq - \frac{h^2}{12} f_{i+\frac{1}{2}}. \end{aligned} \qquad (5.236)$$

Then,

$$
\begin{aligned}
b_3^{(i+1)} &= -\int_{x_i}^{x_{i+1}} f\varphi_{2i+1}\,dx = -\int_{x_i}^{x_{i+1}} \frac{1}{h^3}(x-x_i)^2[2(x_{i+1}-x)+h]f(x)\,dx \\
&\simeq -\frac{h}{6h^3}\left(4\left(x_{i+\frac{1}{2}}-x_i\right)^2\left[2\left(x_{i+1}-x_{i+\frac{1}{2}}\right)+h\right]f_{i+\frac{1}{2}}+h^3 f_{i+1}\right) \\
&\simeq -\frac{h}{6}\left(f_{i+1}+2f_{i+\frac{1}{2}}\right) .
\end{aligned}
\tag{5.237}
$$

Finally, the last component $b_4^{(i+1)}$ is evaluated as:

$$
\begin{aligned}
b_4^{(i+1)} &= -\int_{x_i}^{x_{i+1}} f\varphi_{2i+2}\,dx = -\int_{x_i}^{x_{i+1}} \frac{1}{h^2}(x-x_i)^2(x-x_{i+1})f(x)\,dx \\
&\simeq -\frac{h}{6h^2}\left((0\times f_i)+4\left(x_{i+\frac{1}{2}}-x_i\right)^2\left(x_{i+\frac{1}{2}}-x_{i+1}\right)f_{i+\frac{1}{2}}+(0\times f_{i+1})\right) \\
&\simeq \frac{h^2}{12}f_{i+\frac{1}{2}} .
\end{aligned}
\tag{5.238}
$$

To conclude, the approximation of the elementary vector $b^{(i+1)}$ is written as:

$$
{}^t b^{(i+1)} \simeq \left[-\frac{h}{6}\left(f_i+2f_{i+\frac{1}{2}}\right),\ -\frac{h^2}{12}f_{i+\frac{1}{2}},\ -\frac{h}{6}\left(f_{i+1}+2f_{i+\frac{1}{2}}\right),\ \frac{h^2}{12}f_{i+\frac{1}{2}}\right].
\tag{5.239}
$$

b) *Particular Case of a Uniform Load having Intensity $f_2 \equiv p$.*

When the density f_2 is constant along beam Ω, the elementary vector $b^{(i+1)}$ is written as:

$$
{}^t b_*^{(i+1)} = -p\left[\int_{x_i}^{x_{i+1}}\varphi_{2i-1}\,dx,\ \int_{x_i}^{x_{i+1}}\varphi_{2i}\,dx,\ \int_{x_i}^{x_{i+1}}\varphi_{2i+1}\,dx,\ \int_{x_i}^{x_{i+1}}\varphi_{2i+2}\,dx\right].
\tag{5.240}
$$

Thus, given that the basis functions of $\tilde{W}$ are third degree polynomials, the Simpson quadrature formula is exact for such polynomials and the approximation proposed in (5.239) becomes an exact evaluation provided that values f_i, $f_{i+\frac{1}{2}}$ and f_{i+1} are replaced by the loading constant p.

The following is finally obtained:

$$
{}^t b_*^{(i+1)} = \left[-\frac{ph}{2},\ -\frac{ph^2}{12},\ -\frac{ph}{2},\ \frac{ph^2}{12}\right].
\tag{5.241}
$$

A.11) Now, suppose that the beam is subdivided into three equal meshes having length h. Then x_0, x_1, x_2 and x_3 are the four nodes of the mesh resulting from this three-meshed discretisation.

Before proceeding to the assembly of matrix A and of the second member corresponding to the linear system of the minimization problem $\widetilde{(\mathbf{MP})}$, observe the degree of freedom corresponding to the approximation framework using Hermite's finite elements.

To achieve this, first of all note that given the clamped-clamped beam structure when $x_0 = 0$ and when $x_3 = L$, only the nodes at abscissas x_1 and x_2 constitute the degrees of freedom noted: $\tilde{u}_1, \mathring{u}_1, \tilde{u}_2$ and $\mathring{u}_2$.

Thus, the approximation of the displacements field along beam Ω is written as:

$$\tilde{u} = \tilde{u}_1 \varphi_1 + \mathring{u}_1 \varphi_2 + \tilde{u}_2 \varphi_3 + \mathring{u}_2 \varphi_4 . \tag{5.242}$$

c) *Assembly of global matrix A and that of the second member b corresponding to the $\widetilde{(\mathbf{MP})}$ problem.*

The assembly is performed according to the method set out in problem 5.1.1 by first considering the first element $[x_0, x_1]$.

In this case, only the node at abscissa x_1 makes a contribution to the global matrix A.

In other words, if the elementary matrix $a^{(1)}$ relative to the element $[x_0, x_1]$ is considered, it is the sub-matrix corresponding to the degree of freedom of node x_1, $(\tilde{u}_1, \mathring{u}_1)$ that should be taken into consideration in the assembly of matrix A as follows.

Elements of the elementary matrix $a^{(1)}$ (see definition (5.197)) that should be taken into consideration for the assembly are indicated in bold character as follows:

$$a^{(1)} = \frac{2EI}{h^3} \begin{array}{c} \quad\quad\quad \overset{\tilde{u}_1}{\downarrow} \quad \overset{\mathring{u}_1}{\downarrow} \\ \begin{bmatrix} 6 & 3h & -6 & 3h \\ 3h & 2h^2 & -3h & h^2 \\ -6 & -3h & \mathbf{6} & \mathbf{-3h} \\ 3h & h^2 & \mathbf{-3h} & \mathbf{2h^2} \end{bmatrix} \begin{array}{l} \\ \\ \leftarrow \tilde{u}_1 \\ \leftarrow \mathring{u}_1 \end{array} \end{array} . \tag{5.243}$$

Then, after taking into account the contribution of element $[x_0, x_1]$, the global matrix A has the following form:

$$A = \frac{2EI}{h^3} \begin{array}{cccc} \tilde{u}_1 & \dot{\tilde{u}}_1 & \tilde{u}_2 & \dot{\tilde{u}}_2 \\ \downarrow & \downarrow & & \\ \begin{bmatrix} 6 & -3h & 0 & 0 \\ -3h & 2h^2 & 0 & 0 \\ 0 & 0 & 0 & 0 \\ 0 & 0 & 0 & 0 \end{bmatrix} & \begin{array}{l} \leftarrow \tilde{u}_1 \\ \leftarrow \dot{\tilde{u}}_1 \\ \tilde{u}_2 \\ \dot{\tilde{u}}_2 \end{array} \end{array}$$

(5.244)

Likewise, the contribution of elementary vector $b^{(1)}$, according to the general definition (5.198), corresponds only to the two components $b_3^{(1)}$ and $b_4^{(1)}$ relative to the two degrees of freedom $\tilde{u}_1$ and $\dot{\tilde{u}}_1$.

Thus, after integrating the contribution of the first mesh $[x_0, x_1]$, the second member b is written as:

$$^tb = \left[b_3^{(1)}, b_4^{(1)}, 0, 0 \right].$$

(5.245)

Now, the second mesh $[x_1, x_2]$ of the beam Ω is considered.

In this case, the two nodes at abscissas x_1 and x_2 respectively constitute the degrees of freedom $(\tilde{u}_1, \dot{\tilde{u}}_1)$ and $(\tilde{u}_2, \dot{\tilde{u}}_2)$.

Thus, the elementary matrix $a^{(2)}$, relative to element $[x_1, x_2]$ is full and is written according to definition (5.198):

$$a^{(2)} = \frac{2EI}{h^3} \begin{array}{cccc} \tilde{u}_1 & \dot{\tilde{u}}_1 & \tilde{u}_2 & \dot{\tilde{u}}_2 \\ \downarrow & \downarrow & \downarrow & \downarrow \\ \begin{bmatrix} 6 & 3h & -6 & 3h \\ 3h & 2h^2 & -3h & h^2 \\ -6 & -3h & 6 & -3h \\ 3h & h^2 & -3h & 2h^2 \end{bmatrix} & \begin{array}{l} \leftarrow \tilde{u}_1 \\ \leftarrow \dot{\tilde{u}}_1 \\ \leftarrow \tilde{u}_2 \\ \leftarrow \dot{\tilde{u}}_2 \end{array} \end{array}$$

(5.246)

Then, carry out the assembly by integrating the contribution of matrix $a^{(2)}$ in the global matrix A. This contribution is shown in bold characters in matrix A, (the terms from the elementary matrix $a^{(1)}$ being neutralized in normal font size):

$$
A = \frac{2EI}{h^3}
\begin{array}{cccc}
\tilde{u}_1 & \dot{\tilde{u}}_1 & \tilde{u}_2 & \dot{\tilde{u}}_2 \\
\downarrow & \downarrow & \downarrow & \downarrow \\
\end{array}
\begin{bmatrix}
6+6 & -3h+3h & -6 & 3h \\
-3h+3h & 2h^2+2h^2 & 3h & h^2 \\
-6 & -3h & 6 & -3h \\
3h & h^2 & -3h & 2h^2
\end{bmatrix}
\begin{array}{l}
\leftarrow \tilde{u}_1 \\
\leftarrow \dot{\tilde{u}}_1 \\
\leftarrow \tilde{u}_2 \\
\leftarrow \dot{\tilde{u}}_2
\end{array}
\tag{5.247}
$$

The elementary calculations intervening in the coefficients of the matrix A being performed, the following is obtained:

$$
A = \frac{2EI}{h^3}
\begin{bmatrix}
12 & 0 & -6 & 3h \\
0 & 4h^2 & -3h & h^2 \\
-6 & -3h & 6 & -3h \\
3h & h^2 & -3h & 2h^2
\end{bmatrix}.
\tag{5.248}
$$

Concerning the contribution of mesh $[x_1, x_2]$ in the second member b, the elementary vector $b^{(2)}$ is once again full and is written as:

$$
{}^t b^{(2)} = \left[b_1^{(2)}, b_2^{(2)}, b_3^{(2)}, b_4^{(2)} \right].
\tag{5.249}
$$

Then, the assembly in the second member b gives:

$$
b =
\begin{bmatrix}
b_3^{(1)} + b_1^{(2)} \\
b_4^{(1)} + b_2^{(2)} \\
b_3^{(2)} \\
b_4^{(2)}
\end{bmatrix}.
\tag{5.250}
$$

End the assembly of matrix A by taking into account the contribution of the last mesh $[x_2, x_3]$.

In the present case, there is a situation of symmetry in relation to the one presented for mesh $[x_0, x_1]$.

In fact, in the present case, it is node x_3 that is restrained and only the degrees of freedom of the node at abscissa x_2 i. e. $(\tilde{u}_2, \dot{\tilde{u}}_2)$ should be taken into consideration.

The coefficients of the elementary matrix $a^{(3)}$ to be taken into account are those of the sub-matrix indicated as follows:

$$
a^{(3)} = \frac{2EI}{h^3}
\begin{array}{c}
\quad\;\; \tilde{u}_2 \;\; \dot{\tilde{u}}_2 \\
\quad\;\; \downarrow \;\; \downarrow \\
\left[
\begin{array}{cccc}
6 & 3h & -6 & 3h \\
3h & 2h^2 & -3h & h^2 \\
-6 & -3h & 6 & -3h \\
3h & h^2 & -3h & 2h^2
\end{array}
\right]
\begin{array}{l}
\leftarrow \tilde{u}_2 \\
\leftarrow \dot{\tilde{u}}_2 \\
\\
\end{array}
\end{array}
\tag{5.251}
$$

Then, the assembly in the global matrix gives:

$$
A = \frac{2EI}{h^3}
\begin{array}{c}
\;\; \tilde{u}_1 \quad\;\; \dot{\tilde{u}}_1 \quad\;\;\; \tilde{u}_2 \quad\;\;\; \dot{\tilde{u}}_2 \\
\quad\quad\quad\;\; \downarrow \quad\quad\;\; \downarrow \\
\left[
\begin{array}{cccc}
12 & 0 & -6 & 3h \\
0 & 4h^2 & -3h & h^2 \\
-6 & -3h & 6+6 & -3h+3h \\
3h & h^2 & -3h+3h & 2h^2+2h^2
\end{array}
\right]
\begin{array}{l}
\tilde{u}_1 \\
\dot{\tilde{u}}_1 \\
\leftarrow \tilde{u}_2 \\
\leftarrow \dot{\tilde{u}}_2
\end{array}
\end{array}
\tag{5.252}
$$

Finally, the final form of the matrix corresponding to a mesh of the beam constituted by three meshes is:

$$
A = \frac{2EI}{h^3}
\left[
\begin{array}{cccc}
12 & 0 & -6 & 3h \\
0 & 4h^2 & -3h & h^2 \\
-6 & -3h & 12 & 0 \\
3h & h^2 & 0 & 4h^2
\end{array}
\right].
\tag{5.253}
$$

Given the restraint when $x_3 = L$, the contribution of mesh $[x_2, x_3]$ in the second member intervenes by the elementary vector $b^{(3)}$, and this occurs only through its components $b_1^{(3)}$ and $b_2^{(3)}$.

The final assembly in the second member b then gives:

$$b = \begin{bmatrix} b_3^{(1)} + b_1^{(2)} \\[2mm] b_4^{(1)} + b_2^{(2)} \\[2mm] b_3^{(2)} + b_1^{(3)} \\[2mm] b_4^{(2)} + b_2^{(3)} \end{bmatrix} \simeq \begin{bmatrix} -\dfrac{h}{6}\left(2f_{\frac{1}{2}} + 2f_1 + 2f_{\frac{3}{2}}\right) \\[3mm] \dfrac{h^2}{12}\left(f_{\frac{1}{2}} - f_{\frac{3}{2}}\right) \\[3mm] -\dfrac{h}{6}\left(2f_{\frac{3}{2}} + 2f_2 + 2f_{\frac{5}{2}}\right) \\[3mm] \dfrac{h^2}{12}\left(f_{\frac{3}{2}} - f_{\frac{5}{2}}\right) \end{bmatrix}, \tag{5.254}$$

where the generic result of approximation (5.239) of the elementary vector $b^{(i+1)}$ would have been used.

d) *Approximation of the displacement field at any point of beam* Ω.

In order to find an approximation of the displacement field $\tilde{u}$ at any point of beam Ω, it is only necessary to solve the linear system defined by matrix A (5.253) and by the second member b (5.254).

The corresponding analytical solution produced by a computational solver is given by:

$$\tilde{u}_1 = \frac{1}{171h}\left[(11b_3 + 24b_1)h - 17b_4 - 4b_2\right], \tag{5.255}$$

$$\dot{\tilde{u}}_1 = \frac{1}{57h^2}\left[(5b_3 + 4b_1)h - 6b_4 + 12b_2\right], \tag{5.256}$$

$$\tilde{u}_2 = \frac{1}{171h}\left[(16b_3 + 9b_1)h - 4b_4 - 11b_2\right], \tag{5.257}$$

$$\dot{\tilde{u}}_2 = -\frac{1}{57h^2}\left[(4b_3 + 7b_1)h - 20b_4 + 2b_2\right], \tag{5.258}$$

where b_1, b_2, b_3 and b_4 are the components of the second member b whose approximation has been proposed in (5.254).

At any abscissa x of the beam Ω, the displacement field $\tilde{u}$ is then defined by the formula (5.218) after adapting it to a mesh having three elements, namely:

$$\tilde{u}(x) = \tilde{u}_1\varphi_1(x) + \dot{\tilde{u}}_1\varphi_2(x) + \tilde{u}_2\varphi_3(x) + \dot{\tilde{u}}_2\varphi_4(x), \tag{5.259}$$

where the coefficients $\tilde{u}_1, \dot{\tilde{u}}_1, \tilde{u}_2$ and $\dot{\tilde{u}}_2$ are given by (5.255)–(5.258) and the functions $(\varphi_i)_{i=1,4}$ are the basis functions corresponding to expressions (5.227)–(5.230).

e) *Hermite's finite elements applied to the variational formulation (VP).*

The variational formulation (**VP**) is written according to formula (5.186). In order to obtain the approximate variational formulation ($\widetilde{\textbf{PV}}$), the usual substitutions are performed:

$$u(x) \rightarrow \tilde{u}(x) = \sum_{k=1}^{N} \dot{\tilde{u}}_k \varphi_{2k}(x) + \sum_{k=1}^{N} \tilde{u}_k \varphi_{2k-1}(x) , \tag{5.260}$$

$$v(x) \rightarrow \tilde{v}(x) = \varphi_i(x) . \tag{5.261}$$

The approximate formulation ($\widetilde{\textbf{VP}}$) is then written as:

$$(\widetilde{\textbf{VP}}) \begin{bmatrix} \text{Find } \tilde{u} \text{ solution to:} \\[2mm] EI \sum_{k=1}^{N} \int_0^L \left[\dot{\tilde{u}}_k \varphi_{2k} + \tilde{u}_k \varphi_{2k-1} \right] \varphi_i \, dx = - \int_0^L f \varphi_i \, dx , \\[2mm] i = 1 \text{ to } 2N . \end{bmatrix} \tag{5.262}$$

Then, it is only necessary to note that the $2N$ equations parameterised by i in formulation (5.262) may be split into two groups: those corresponding to the even values of i and those corresponding to the odd values of i.

This distinction then strictly yields the same formulation as that of the system of equations (5.233)–(5.234) of the minimization problem ($\widetilde{\textbf{MP}}$).

Wherefrom, it is inferred that, in the case of a discretisation with three meshes, the nodal equations of the approximate variational formulation ($\widetilde{\textbf{VP}}$) resulting from it leads to the same linear system of matrix A defined by (5.253) and of the second member b defined by (5.254).

Given the properties of the bilinear form $a(.,.)$ and of the linear form $L(.)$ defined by (5.215), this result is nothing but the consequence of the equivalence between the variational problem (**VP**) and the minimization problem (**MP**).

Chapter 6

Finite Elements Applied to Non Linear Problems

6.1 Viscous Burgers Equation

▶ **Warning**

This problem deals with the viscous Burgers equation as an approximation to the Navier-Stokes equation in the one-dimensional case.

In order to suggest the study of a "Finite elements in space – Finite differences in time" mixed formulation that may be accessible to science graduate students who do not possess all the knowledge in functional analysis necessary for processing the result, all that is inherent to the definition of the functional framework has been voluntarily excluded in the subsequent presentation.

In other words, only the formal aspects of the variational formulations and of the numerical application of the finite elements are considered in all that will follow.

Statement

1) Here, the scalar function u of variables (x,t) is of interest as solution to the following partial differential equation:

$$(CP) \begin{cases} \dfrac{\partial u}{\partial t} + u\dfrac{\partial u}{\partial x} = v\dfrac{\partial^2 u}{\partial x^2}, & \forall (x,t) \in]0,L[\times]0,+\infty[, \\[2mm] u(0,t) = 0, \quad \dfrac{\partial u}{\partial x}(L,t) = f(t), & \forall t \in]0,+\infty[, \\[2mm] u(x,0) = u_0(x), & \forall x \in]0,L[, \end{cases} \tag{6.1}$$

where v denotes the kinematic viscosity of the fluid, L a given and characteristic length of the flow and f, a "sufficiently regular" function is also given.

– What is the fundamental property of the partial differential equation (**CP**)?

2) Let v be a test function of the only variable x. Show that the (**CP**) problem can be expressed in the following variational formulation (**VP**):

Find u belonging to V solution of:

$$(\textbf{VP})\frac{\mathrm{d}}{\mathrm{d}t}\int_0^L uv\,\mathrm{d}x + v\int_0^L \frac{\partial u}{\partial x}\frac{\mathrm{d}v}{\mathrm{d}x}\,\mathrm{d}x + \int_0^L u\frac{\partial u}{\partial x}v\,\mathrm{d}x = vf(t)v(L)\,, \quad \forall v \in V\,. \quad (6.2)$$

The boundary conditions satisfying the functions v of V are to be specified without discussing their characteristics for functional regularity.

3) The approximation of the variational problem (**VP**) is done using Lagrange finite elements P_1.

To achieve this, a regular mesh of constant step h is introduced at interval $[0,L]$, such that:

$$\begin{cases} x_0 = 0,\ x_{N+1} = L\,, \\ x_{i+1} = x_i + h,\ i = 0 \text{ to } N\,. \end{cases} \quad (6.3)$$

In addition, the approximation space $\tilde{V}$ is defined by:

$$\tilde{V} = \left\{\tilde{v}\colon [0,L] \to \mathbf{R},\ \tilde{v} \in C^o([0,1]),\ \tilde{v}|_{[x_i,x_{i+1}]} \in P_1,\ \tilde{v}(0) = 0\right\}\,, \quad (6.4)$$

where $P_1\,([x_i,x_{i+1}])$ denotes the polynomial space defined on $[x_i,x_{i+1}]$, of degree less than or equal to one.

– What is the dimension of $\tilde{V}$?

4) Let φ_i, $(i = 1 \text{ to } \dim\tilde{V})$ be the basis of $\tilde{V}$ satisfying $\varphi_i(x_j) = \delta_{ij}$.

After expressing approximated variational formulation of solution $\tilde{u}$ associated to the variational problem (**VP**), show that when choosing:

$$\tilde{v}(x) = \varphi_i(x) \quad \text{and} \quad \tilde{u}(x,t) = \sum_{j=1,\,\dim\tilde{V}} \tilde{u}_j(t)\varphi_j(x)\,, \quad (6.5)$$

the differential system is obtained in time (**DS**):

$$\begin{array}{|l}
A_{ij}\tilde{u}'_j(t) + B_{ij}\tilde{u}_j(t) + C_{ijk}\tilde{u}_j(t)\tilde{u}_k(t) = \tilde{F}_i(t)\,, \\[4pt]
\forall i \in \{1,\dots,\dim\tilde{V}\}\,, \quad \forall t \geq 0\,. \hfill (6.6) \\[4pt]
\text{where it was stated:} \\[4pt]
(\textbf{DS}) \quad \tilde{u}'_j(t) = \dfrac{\mathrm{d}}{\mathrm{d}t}\tilde{u}_j(t),\ A_{ij} = \displaystyle\int_0^L \varphi_i\varphi_j\,\mathrm{d}x\,, \\[8pt]
\qquad\qquad B_{ij} = v\displaystyle\int_0^L \frac{\mathrm{d}\varphi_i}{\mathrm{d}x}\frac{\mathrm{d}\varphi_j}{\mathrm{d}x}\,\mathrm{d}x\,, \hfill (6.7) \\[8pt]
\qquad\qquad C_{ijk} = \displaystyle\int_0^L \varphi_i\varphi_j\frac{\mathrm{d}\varphi_k}{\mathrm{d}x}\,\mathrm{d}x,\ \tilde{F}_i(t) = vf(t)\varphi_i(L)\,. \hfill (6.8)
\end{array}$$

In addition, the repeated indices summation convention (or Einstein convention) would have been used where an index repeated in the same monomial is to be summed over all possible values that this index can take:

$$X_j Y_j \equiv \sum_{j=1,\,\dim \hat{V}} X_j Y_j \,. \tag{6.9}$$

– What is the major characteristic of the differential system (DS)?

▶ **Basis Function φ_i Characteristic of a Node Strictly Interior at $[0,L]$**

5) Considering the regularity of the mesh, the generic equation of the system (DS) associated to any basis function φ_i, characteristic of a node strictly interior at $[0,L]$, is expressed as:

$$(\mathbf{DS_1}) \begin{cases} A_{i,i-1}\,\tilde{u}'_{i-1}(t) + A_{i,i}\,\tilde{u}'_i(t) + A_{i,i+1}\,\tilde{u}'_{i+1}(t) + \ldots \\[4pt] B_{i,i-1}\,\tilde{u}_{i-1}(t) + B_{i,i}\,\tilde{u}_i(t) + B_{i,i+1}\,\tilde{u}_{i+1}(t) + \ldots \\[4pt] C_{i,i-1,i-1}\,\tilde{u}^2_{i-1}(t) + C_{i,i-1,i}\,\tilde{u}_{i-1}(t)\tilde{u}_i(t) + C_{i,i,i-1}\,\tilde{u}_i(t)\tilde{u}_{i-1}(t) + \ldots \quad (6.10) \\[4pt] C_{i,i,i}\,\tilde{u}^2_i(t) + C_{i,i,i+1}\,\tilde{u}_i(t)\tilde{u}_{i+1}(t) + C_{i,i+1,i}\,\tilde{u}_{i+1}(t)\tilde{u}_i(t) + \ldots \\[4pt] C_{i,i+1,i+1}\,\tilde{u}^2_{i+1}(t) = 0 \,. \end{cases}$$

– Using the trapezium rule, calculate the 13 coefficients $(A_{ij}, B_{ij}, C_{ijk})$ of system (**DS$_1$**) and express the corresponding nodal equation.

– Show that the scheme of centred finite differences associated to the differential equation of the continuous problem (**CP**) is found. What is its order?

It is reminded that the trapezium quadrature formula is expressed as:

$$\int_a^b \xi(s)\,ds \simeq \frac{(b-a)}{2}\{\xi(a) + \xi(b)\} \,.$$

▶ **Basis Function φ_{N+1} Characteristic of the Abscissa Node $x_{N+1} = L$**

6) The same process is used for the basis function φ_{N+1}, characteristic of the final node x_{N+1}. The corresponding equation to the system (**DS**) is then expressed as:

$$(\mathbf{DS_2}) \begin{cases} A_{N+1,N}\,\tilde{u}'_N(t) + A_{N+1,N+1}\,\tilde{u}'_{N+1}(t) + \ldots \\[4pt] B_{N+1,N}\,\tilde{u}_N(t) + B_{N+1,N+1}\,\tilde{u}_{N+1}(t) + \ldots \\[4pt] C_{N+1,N+1,N+1}\,\tilde{u}^2_{N+1}(t) + C_{N+1,N,N+1}\,\tilde{u}_N(t)\tilde{u}_{N+1}(t) + \ldots \\[4pt] C_{N+1,N,N}\,\tilde{u}^2_N(t) + C_{N+1,N+1,N}\,\tilde{u}_N(t)\tilde{u}_{N+1}(t) = \nu\,f(t). \end{cases} \tag{6.11}$$

– Using the trapezium rule, calculate the 8 coefficients $(A_{ij}, B_{ij}, C_{ijk})$ of equation (DS_2) and express the corresponding nodal equation.

– Show that the scheme of finite differences of the second order is found by discretising the Neumann boundary condition of the problem (CP).

▶ Discretisation by Finite Differences in Time

7) Suggest a discretisation by finite differences in time of the differential system $(DS_1)-(DS_2)$.

6.2 Solution

A.1) Of course, the major characteristic of the Burgers equation of the continuous problem **(CP)** is its inherent non-linearity to the matching advection-convection term $u\dfrac{\partial u}{\partial x}$.

A number of research works have enabled the exploration of the properties of this equation as being the particular case of a non-linear hyperbolic partial differential equation.

The interested reader may consult the work of D. Euvrard [4] for an elementary presentation intended for mechanics or physics graduate students. The work of Edwige Godlewski and Pierre-Arnaud Raviart provide further in depth studies requiring a good command of the basic techniques in functional analysis [6].

A.2) Now, it is proposed to find a variational formulation leading to the application of the Lagrange finite elements P_1 in space.

To achieve this, consider test functions v, defined on $[0, L]$ and having real values. In other words, test functions v are a function of the *only* space variable x.

Then, the equation with partial derivatives of the continuous problem **(CP)** defined in (6.1) is multiplied by v and integrated over the interval $[0, L]$.

$$\int_0^L \frac{\partial u}{\partial t} v(x)\,\mathrm{d}x + \int_0^L u\frac{\partial u}{\partial x} v(x)\,\mathrm{d}x = v\int_0^L \frac{\partial^2 u}{\partial x^2} v(x)\,\mathrm{d}x. \tag{6.12}$$

Now, the second member of (6.12) is integrated by parts and the result is:

$$\int_0^L \frac{\partial u}{\partial t} v\,\mathrm{d}x + \int_0^L u\frac{\partial u}{\partial x} v\,\mathrm{d}x = -v\int_0^L \frac{\partial u}{\partial x}\frac{\mathrm{d}v}{\mathrm{d}x}\,\mathrm{d}x\ldots$$
$$+ v\left[\frac{\partial u}{\partial x}(L,t)\cdot v(L) - \frac{\partial u}{\partial x}(0,t)\cdot v(0)\right]. \tag{6.13}$$

By using boundary conditions of the problem **(CP)** when $x = L$ on one hand and by requiring test functions v to be zero when $x = 0$ on the other hand, in order to keep the wholeness of the "in space" information of the formulation of the continuous problem, the following variational formulation **(VP)** is obtained:

$$\textbf{(VP)}\quad \left[\begin{array}{l} \text{Find } u \text{ belonging to } V \text{ solution to:} \\[4pt] \displaystyle\int_0^L \frac{\partial u}{\partial t} v\,\mathrm{d}x + \int_0^L u\frac{\partial u}{\partial x} v\,\mathrm{d}x = -v\int_0^L \frac{\partial u}{\partial x}\frac{\mathrm{d}v}{\mathrm{d}x}\,\mathrm{d}x + v f(t)\cdot v(L), \\[8pt] \forall v / v(0) = 0. \end{array}\right. \tag{6.14}$$

It would be noticed that the variational formulation in space **(VP)** is only formal, insofar as the functional framework V, in which this formulation makes sense, was completely omitted.

For further details, the reader accustomed to the basic techniques of functional analysis, namely to the Sobolev spaces $H^m[0,T;L^p(\mathbf{R})]$ may consult the work of Edwige Godlewski and Pierre Arnaud Raviart [6] mentioned above.

R.3) Estimation of the dimension of approximation space $\tilde{V}$ is carried out as follows.

The functions belonging to space are exactly defined by the $(N+1)$ values at nodes $(x_i)_{i=1,N+1}$ in relation to the mesh (6.3).

It would be noticed that the node at abscissa $x_0 = 0$ does not contribute any degree of freedom insofar as the value of any function $\tilde{v}$ of $\tilde{V}$ is zero at this point.

To ascertain that, it is only necessary to proceed to the visualization of such functions. In fact, any function of $\tilde{v}$ the approximation space $\tilde{V}$ is a pecked line formed by affine functions per mesh $[x_i,x_{i+1}]$ and whose interior nodes $x_i, (i = 1$ to $N)$ constitute the points of continuity between the two adjacent mesh.

That is why, only the values at $(N+1)$ nodes $(x_1,x_2,\ldots,x_{N+1})$ display a degree of freedom for any function $\tilde{V}$ of $\tilde{v}$. Changing of one of these $(N+1)$ values immediately requires a modification of the given element of into another function $\tilde{v}$ of $\tilde{V}$.

Thus, without any formal demonstration, it is observed that understanding a function $\tilde{v}$ of $\tilde{V}$ is equivalent to the data of a vector of $\mathbf{R}^{N+1}$ constituted the $(N+1)$ values $(\tilde{v}_1,\ldots,\tilde{v}_{N+1})$ of any function $\tilde{v}$ of $\tilde{V}$.

In other words, the dimension of $\tilde{V}$ is equal to $(N+1)$ insofar as this space is isomorphic at $\mathbf{R}^{N+1}$.

R.4) The approximate variational formulation is obtained by substituting the functions (u,v) of functional space V by their respective approximations $(\tilde{u},\tilde{v})$ belonging to space $\tilde{V}$.

Thus, the approximate variational formulation $\widetilde{(\mathbf{VP})}$ is written as:

$$\widetilde{(\mathbf{VP})} \quad \left[\begin{array}{l} \text{Find } \tilde{u} \text{ belonging to } \tilde{V} \text{ solution to:} \\[2mm] \dfrac{\mathrm{d}}{\mathrm{d}t}\displaystyle\int_0^L \tilde{u}\tilde{v}\,\mathrm{d}x + v\int_0^L \dfrac{\partial \tilde{u}}{\partial x}\dfrac{\mathrm{d}\tilde{v}}{\mathrm{d}x}\,\mathrm{d}x + \int_0^L \tilde{u}\dfrac{\partial \tilde{u}}{\partial x}\tilde{v}\,\mathrm{d}x = vf(t)\tilde{v}(L)\,, \\[2mm] \forall \tilde{v} \in \tilde{V}\,. \end{array} \right. \qquad (6.15)$$

Now, a particular case in which functions $\tilde{v}$ are basis functions φ_i of space $\tilde{V}$ is examined and the development of the approximate solution $\tilde{u}$ is carried out on this basis:

$$\tilde{v}(x) = \varphi_i(x) \quad \text{and} \quad \tilde{u}(x,t) = \sum_{j=1,N+1} \tilde{u}_j(t)\varphi_j(x)\,. \qquad (6.16)$$

Given that the non-linearity is inherent to the advection-convection term $\tilde{u}\dfrac{\partial \tilde{u}}{\partial x}\tilde{v}$, it is necessary to consider the development of approximation $\tilde{u}$ (previously proposed in (6.16)), by introducing two summation indices j and k as follows.

Then, the approximate variational formulation $\widetilde{(\mathbf{VP})}$ defined by (6.15) can then be rewritten in the form:

$$\tilde{u}'_j(t)\int_0^L \varphi_i\varphi_j\,dx + \left[v\int_0^L \frac{d\varphi_i}{dx}\frac{d\varphi_j}{dx}\,dx\right]\tilde{u}_j(t) + \left[\int_0^L \varphi_i\varphi_j\frac{d\varphi_k}{dx}\,dx\right]\tilde{u}_j(t)\tilde{u}_k(t)$$
$$= vf(t)\varphi_i(L)\,, \quad (\forall i = 1 \text{ to } N+1)\,, \tag{6.17}$$

where the convention of repeated indices (6.9) is adopted.

Then, by introducing notations (6.7)–(6.8), the variational approximation (6.17) produces in time the differential system (**DS**) defined by:

$$A_{ij}\tilde{u}'_j(t) + B_{ij}\tilde{u}_j(t) + C_{ijk}\tilde{u}_j(t)\tilde{u}_k(t) = \tilde{F}_i(t)\,, \quad (\forall i = 1 \text{ to } N+1)\,, \forall t \geq 0\,. \tag{6.18}$$

Of course, insofar as the Burgers equation of the continuous problem (**CP**) defined by (6.1) is non-linear, this characteristic is omnipresent in the variational formulation (6.2) as well as in its approximate form (6.15) and, as a result, in the non-linear system (**DS**).

This non-linearity would then require appropriate numerical methods like the Newton's method, (cf. [2]) in order to produce an approximation of the differential system (**DS**).

R.5) Now, a characteristic basis function φ_i, $(i = 1 \text{ to } N)$ of a node strictly interior at the mesh defined by (6.3) is considered.

For each of these basis functions φ_i, insofar as its support consists of the union of the two intervals $[x_{i-1}, x_i]$ and $[x_i, x_{i+1}]$ (cf. Fig 6.1), the coefficients of A_{ij} and B_{ij} matrices and those defined by C_{ijk} (cf. (6.7)–(6.8)) will produce non-zero terms if and only if the supports of functions φ_j and φ_k have a non-vacuous intersection with that of the considered function φ_i.

In order to find the appropriate terms to be considered for the formation of the differential system (**DS**), the index i, $(i = 1$, to $N)$ is fixed.

In this case, concerning the A_{ij} and B_{ij} matrices, it immediately comes out that only basis functions φ_{i-1}, φ_i and φ_{i+1} and can produce a non-zero integration against the basis function φ_i, (cf. Fig. 6.1).

That is why, only coefficients A_{ij}, $(j = i-1, i, i+1)$ and B_{ij}, $(j = i-1, i, i+1)$ are to be retained while writing the differential system (**DS**).

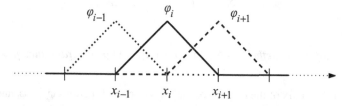

Fig. 6.1 Basis Functions φ_{i-1}, φ_i and φ_i

Now, the term C_{ijk} whose contributions may be subject to a finer analysis is considered.

Having fixed $i, (i = 1$ to $N)$ because of the structure of coefficient C_{ijk}, it is only necessary to consider the following three values of j as done previously: $i-1, i$ and $i+1$.

Thus, the following pairs of indices to be considered are already available: $(i, i-1)$, (i, i) and $(i, i+1)$.

Then, the values of index k are determined for each of these pairs – those likely to produce non-zero terms in the non-linear system (**DS**).

Concerning the pair $(i, i-1)$, only the values of k equal to $i-1$ and i are to be retained. In fact, the value $k = i+1$ would lead to a coefficient $C_{i,i-1,i+1}$ that would be zero insofar as the supports of basis function φ_{i-1} are separate from that of function φ_{i+1}.

Thus, the triplet of indices $(i, i-1, i-1)$ and $(i, i-1, i)$ needs to be considered.

Likewise, the pair (i, i) requires consideration of the following triplet of indices: $(i, i, i-1)$, (i, i, i) and $(i, i, i+1)$.

Finally, pair $(i, i+1)$ is the counterpart of pair $(i, i-1)$ and the following triplets will be considered: $(i, i+1, i)$ and $(i, i+1, i+1)$.

All coefficients that can produce non-zero terms in the non-linear system (**DS**) are grouped below:

$$C_{i,i-1,i-1}, \ C_{i,i-1,i}, \ C_{i,i,i-1}, \ C_{i,i,i}, \ C_{i,i,i+1}, \ C_{i,i+1,i}, \ C_{i,i+1,i+1}. \tag{6.19}$$

Thus, if coefficients $A_{i,j}$, $B_{i,j}$ and $C_{i,j,k}$, just identified as part of the definition of the generic equation of the non-linear differential system (**DS**), are retained, the N equations (**DS$_1$**) defined by (6.10) are then obtained.

Moreover, it would have been noticed that the second member $\tilde{F}_i(t) \equiv vf(t)\varphi_i(L)$ is identically zero for these basis functions φ_i, $(i = 1$ to $N)$, given the availability of the property:

$$\varphi_i(L) = \varphi_i(x_N) = 0, \quad \forall i = 1 \text{ to } N. \tag{6.20}$$

At the same time, it will be noticed that all the basis functions φ_i considered here satisfy the property:

$$\varphi_i(x_j) = \delta_{ij}, \quad \forall (i, j) \in \{1, \dots, N+1\}. \tag{6.21}$$

a) *Approximation of coefficients* $A_{ij}, (j = i-1, i, i+1)$ *by the trapezium quadrature formula.*

Given the relative fix of the respective supports of basis functions φ_{i-1}, φ_i and φ_{i+1}, the following is obtained:

$$A_{i,i-1} \equiv \int_0^L \varphi_i \varphi_{i-1}\, dx = \int_{x_{i-1}}^{x_i} \varphi_i \varphi_{i-1}\, dx$$

$$\simeq \frac{h}{2}\left[\varphi_i(x_{i-1})\varphi_{i-1}(x_{i-1}) + \varphi_i(x_i)\varphi_{i-1}(x_i)\right] = 0, \tag{6.22}$$

where the property (6.21) of basis functions φ_i would have been used.

Likewise,

$$A_{i,i+1} \equiv \int_0^L \varphi_i \varphi_{i+1}\, dx = \int_{x_i}^{x_{i+1}} \varphi_i \varphi_{i+1}\, dx$$

$$\simeq \frac{h}{2}\left[\varphi_i(x_i)\varphi_{i+1}(x_i) + \varphi_i(x_{i+1})\varphi_{i+1}(x_{i+1})\right] = 0. \tag{6.23}$$

A classical evaluation of coefficient $A_{i,i}$ is carried out as follows:

$$A_{i,i} = \int_0^L \varphi_i^2\, dx = \int_{x_{i-1}}^{x_i} \varphi_i^2\, dx + \int_{x_i}^{x_{i+1}} \varphi_i^2\, dx$$

$$\simeq \frac{h}{2}[1+0] + \frac{h}{2}[0+1] = h. \tag{6.24}$$

b) *Estimation of coefficients B_{ij}, $(j = i-1, i, i+1)$.*

Calculation of the three coefficients B_{ij} may be performed either by the exact method or by approximation *via* the trapezium quadrature formula, insofar as the latter is exact on the constant functions.

In fact, since functions φ_i are piecewise affine, their derivative is constant per mesh $[x_i, x_{i+1}]$.

Then the following is then obtained:

$$B_{i,i-1} = v \int_0^L \frac{d\varphi_i}{dx}\frac{d\varphi_{i-1}}{dx}\, dx = v \int_{x_{i-1}}^{x_i} \frac{d\varphi_i}{dx}\frac{d\varphi_{i-1}}{dx}\, dx$$

$$= vh \times \left(\frac{1}{h}\right) \times \left(-\frac{1}{h}\right) = -\frac{v}{h}. \tag{6.25}$$

Moreover, the symmetry of B_{ij} matrix on one hand and the invariance of the mesh having constant step h on the other hand make it possible to write:

$$\textbf{Symmetry}\quad\textbf{Invariant}$$
$$\downarrow\qquad\qquad\downarrow$$
$$B_{i,i-1} \;=\; B_{i-1,i} \;=\; B_{i,i+1} \;=\; -\frac{v}{h}. \tag{6.26}$$

Finally, calculation of coefficient $B_{i,i}$ is performed as follows:

$$B_{i,i} = v \int_0^L \left(\frac{d\varphi_i}{dx}\right)^2 dx = v \int_{x_{i-1}}^{x_i} \left(\frac{d\varphi_i}{dx}\right)^2 dx + v \int_{x_i}^{x_{i+1}} \left(\frac{d\varphi_i}{dx}\right)^2 dx$$

$$= vh \left[\frac{1}{h^2} + \left(-\frac{1}{h}\right)^2\right] = \frac{2v}{h} . \tag{6.27}$$

c) *Estimation of coefficients C_{ijk}.*

Starting with a qualitative observation:

Given that coefficients C_{ijk} imply the double product $\varphi_i \times \varphi_j \times \varphi_k'$ the trapezium quadrature formula applied to the calculation of these coefficients will contain only terms of the form:

$$\pm \frac{1}{h} \varphi_i(x_l) \times \varphi_j(x_m) .$$

Moreover, as all the basis functions φ_i satisfy the properties $\varphi_i(x_j) = \delta_{ij}$ when the value of index k is fixed (see the analysis above), the only case to be processed for this value of k corresponds to the one parameterised by $i = j$.

In other words, coefficients $C_{i,i,i-1}, C_{i,i,i}$ and $C_{i,i,i+1}$ are relevant in the calculation, the other four in the collection (6.19) being all identically zero.

In fact, for the other coefficients C_{ijk} that should be considered when, for example, $\varphi_i(x_i)$ is equal to one, $\varphi_{i-1}(x_i)$ and $\varphi_{i+1}(x_i)$ would be identically zero.

It then becomes:

$$C_{i,i,i-1} = \int_0^L (\varphi_i)^2 \cdot \frac{d\varphi_{i-1}}{dx} dx = \int_{x_{i-1}}^{x_i} (\varphi_i)^2 \cdot \frac{d\varphi_{i-1}}{dx} dx$$

$$\simeq -\frac{1}{h} \times \left[\frac{h}{2}(1+0)\right] = -\frac{1}{2} . \tag{6.28}$$

Likewise,

$$C_{i,i,i+1} = \int_0^L (\varphi_i)^2 \cdot \frac{d\varphi_{i+1}}{dx} dx = \int_{x_i}^{x_{i+1}} (\varphi_i)^2 \cdot \frac{d\varphi_{i+1}}{dx} dx ,$$

$$\simeq \frac{1}{h} \times \left[\frac{h}{2}(1+0)\right] = \frac{1}{2} \tag{6.29}$$

Finally, the calculation of coefficient $C_{i,i,i}$ is performed in the following way:

$$C_{i,i,i} = \int_0^L (\varphi_i)^2 \cdot \frac{d\varphi_i}{dx} dx = \int_{x_{i-1}}^{x_i} (\varphi_i)^2 \cdot \frac{d\varphi_i}{dx} dx + \int_{x_i}^{x_{i+1}} (\varphi_i)^2 \cdot \frac{d\varphi_i}{dx} dx$$

$$\simeq \frac{h}{2}\left[\frac{1}{h}(1+0)\right] + \frac{h}{2}\left[-\frac{1}{h}(0+1)\right] = 0 . \tag{6.30}$$

d) *Nodal equation associated with a basis function* φ_i, *characteristic of a node* x_i, $(i = 1 \text{ to } N)$.

The results obtained from (6.19) to (6.30) are grouped in order to write the corresponding nodal equation (6.18):

$$h\tilde{u}_i'(t) - \frac{\nu}{h}[\tilde{u}_{i-1}(t) + \tilde{u}_{i+1}(t)] + \frac{2\nu}{h}\tilde{u}_i(t) - \frac{1}{2}\tilde{u}_i(t)\tilde{u}_{i-1}(t) + \frac{1}{2}\tilde{u}_i(t)\tilde{u}_{i+1}(t) = 0.$$
(6.31)

e) *Finite differences scheme in space associated with the viscous Burgers equation.*

The nodal equation (6.31) is rewritten in the form:

$$\tilde{u}_i'(t) + \tilde{u}_i(t)\left[\frac{\tilde{u}_{i+1}(t) - \tilde{u}_{i-1}(t)}{2h}\right] = \nu\left[\frac{\tilde{u}_{i-1}(t) - 2\tilde{u}_i(t) + \tilde{u}_{i+1}(t)}{h^2}\right].$$
(6.32)

In this form, it is easy to recognize the second order approximation by finite differences of the second partial derivative $\dfrac{\partial^2 u}{\partial x^2}$ at point (x_i, t):

$$\frac{\partial^2 u}{\partial x^2}(x_i, t) = \frac{u(x_{i-1}, t) - 2u(x_i, t) + u(x_{i+1}, t)}{h^2} + O(h^2).$$
(6.33)

Moreover, by performing two Taylor's expansions at point x_i, one progressive and the other regressive to the third order, the following is obtained:

$$u(x_{i+1}, t) = u(x_i, t) + h\frac{\partial u}{\partial x}u(x_i, t) + \frac{h^2}{2}\frac{\partial^2 u}{\partial x^2}(x_i, t) + O(h^3),$$
(6.34)

$$u(x_{i-1}, t) = u(x_i, t) - h\frac{\partial u}{\partial x}u(x_i, t) + \frac{h^2}{2}\frac{\partial^2 u}{\partial x^2}(x_i, t) + O(h^3).$$
(6.35)

Then, the difference between (6.34) and (6.35) results in:

$$\frac{\partial u}{\partial x}(x_i, t) = \frac{u(x_{i+1}, t) - u(x_{i-1}, t)}{2h} + O(h^2).$$
(6.36)

Thus, nodal equation (6.32) is nothing but the second-order approximation by finite differences of the viscous Burgers equation of the continuous problem **(CP)** defined by (6.1).

To ascertain that, it is only necessary to replace the finite differences (6.33) and (6.36) in the Burgers equation of the problem (6.1) to obtain:

$$\frac{\partial u}{\partial t}(x_i, t) + \frac{u(x_{i+1}, t) - u(x_{i-1}, t)}{2h} = \nu\frac{u(x_{i-1}, t) - 2u(x_i, t) + u(x_{i+1}, t)}{h^2} + O(h^2).$$
(6.37)

Then, by eliminating the residue $O(h^2)$ in (6.37) and moving to the approximations in order to maintain equality between the two members of the equation, the nodal equation (6.32) whose approximation is thus of the second order is obtained exactly.

R.6) In order to obtain the equation (**DS₂**), the approximate variational formulation (6.18) is used again and the particular case of the generic basis function φ_{N+1} characteristic of the last node x_{N+1} in relation to the mesh of the $[0,L]$ interval is considered.

Moreover, expansion (6.16) of approximation $\tilde{u}$ in the canonical basis φ_i, $(i = 1$ to $N+1)$ would have been maintained.

The approximate variational equation (6.18) is then written as:

$$A_{N+1,j}\,\tilde{u}'_j(t) + B_{N+1,j}\,\tilde{u}_j(t) + C_{N+1,j,k}\,\tilde{u}_j(t)\tilde{u}_k(t) = \tilde{F}_{N+1}(t)\,, \quad \forall t \geq 0\,, \quad (6.38)$$

by adopting, as usual, the summation of repeated indices convention (6.9).

From then on, the analysis of the coefficients to be considered in equation (6.38), that may produce non-zero contribution, is used again.

Given that the basis function φ_{N+1} consists of a "half cap" (cf. Fig. 6.2), whose support is reduced to the $[x_N, x_{N+1}]$ interval, the only values of index j to be considered are $j = N$ and $j = N+1$ corresponding to functions whose supports intercept that of function φ_{N+1}.

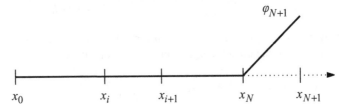

Fig. 6.2 Basis Function φ_{N+1}

In fact, any other value of index j would lead to the consideration of a basis function φ_j whose support would have an empty intersection with that of function φ_{N+1}.

Thus, the coefficients of A_{ij} and B_{ij} matrices to be retained for evaluation are:

$$A_{N+1,N}\,,\ A_{N+1,N+1}\,,\ B_{N+1,N} \quad \text{and} \quad B_{N+1,N+1}\,. \quad (6.39)$$

Likewise, for those two values of index j, only the values of k corresponding to $k = N$ and $k = N+1$ should be considered for the estimation of non-zero coefficients C_{ijk} of equation (6.38) of the non-linear system (**DS**).

The coefficients C_{ijk} to be evaluated are thus:

$$C_{N+1,N,N}\,,\ C_{N+1,N,N+1}\,,\ C_{N+1,N+1,N} \quad \text{and} \quad C_{N+1,N+1,N+1}\,. \quad (6.40)$$

Finally, it is specified that the second member $\tilde{F}_{N+1}(t)$ of equation (6.38) exactly equals:

$$\tilde{F}_{N+1}(t) = vf(t)\varphi_{N+1}(x_{N+1}) = vf(t)\,.$$

a) *Approximation of coefficients $A_{N+1,N}$, $A_{N+1,N+1}$ and $B_{N+1,N+1}$ by the trapezium quadrature formula.*

The calculation of coefficients $A_{N+1,N}$ and $A_{N+1,N+1}$ followed by that of $B_{N+1,N}$ and $B_{N+1,N+1}$ is performed according to the same logic as presented in the answer to question **5**.

Thus, the following is obtained:

$$A_{N+1,N} \equiv \int_0^L \varphi_N \varphi_{N+1}\, dx = \int_{x_N}^{x_{N+1}} \varphi_N \varphi_{N+1}\, dx$$
$$\simeq \frac{h}{2}\left[\varphi_N(x_N)\varphi_{N+1}(x_N) + \varphi_N(x_{N+1})\varphi_{N+1}(x_{N+1})\right] = 0\,, \qquad (6.41)$$

Likewise:

$$A_{N+1,N+1} = \int_0^L (\varphi_{N+1})^2\, dx = \int_{x_N}^{x_{N+1}} (\varphi_{N+1})^2\, dx \simeq \frac{h}{2}[1+0] = \frac{h}{2}\,. \qquad (6.42)$$

Then,

$$B_{N+1,N} = v \int_0^L \frac{d\varphi_{N+1}}{dx}\frac{d\varphi_N}{dx}\, dx = v \int_{x_N}^{x_{N+1}} \frac{d\varphi_{N+1}}{dx}\frac{d\varphi_N}{dx}\, dx$$
$$= vh \times \left(\frac{1}{h}\right) \times \left(-\frac{1}{h}\right) = -\frac{v}{h}\,. \qquad (6.43)$$

Moreover,

$$B_{N+1,N+1} = v \int_0^L \left(\frac{d\varphi_{N+1}}{dx}\right)^2\, dx = v \int_{x_N}^{x_{N+1}} \left(\frac{d\varphi_{N+1}}{dx}\right)^2\, dx$$
$$= vh \times \left(\frac{1}{h}\right)^2 = \frac{v}{h}\,. \qquad (6.44)$$

b) *Approximation of coefficients $C_{N+1,N,N}$, $C_{N+1,N,N+1}$, $C_{N+1,N+1,N}$ and $C_{N+1,N+1,N+1}$ by the trapezium quadrature formula.*

The structural observations, presented in the estimation of coefficients C_{ijk} of question **5** are as valid as in the present case.

That is why only coefficients $C_{N+1,N+1,N}$ and $C_{N+1,N+1,N+1}$ will be estimated, the others being trivially zero.

Then the following is obtained:

$$C_{N+1,N+1,N} = \int_0^L (\varphi_{N+1})^2 \cdot \frac{d\varphi_N}{dx}\, dx = \int_{x_N}^{x_{N+1}} (\varphi_{N+1})^2 \cdot \frac{d\varphi_N}{dx}\, dx\,,$$
$$\simeq -\frac{1}{h}\left[\frac{h}{2}(1+0)\right] = -\frac{1}{2} \qquad (6.45)$$

Likewise:

$$C_{N+1,N+1,N+1} = \int_0^L (\varphi_{N+1})^2 \cdot \frac{d\varphi_{N+1}}{dx} \, dx = \int_{x_N}^{x_{N+1}} (\varphi_{N+1})^2 \cdot \frac{d\varphi_{N+1}}{dx} \, dx \,,$$

$$\simeq \frac{1}{h} \left[\frac{h}{2}(1+0) \right] = \frac{1}{2} \tag{6.46}$$

c) *Nodal equation associated with the basis function* φ_{N+1} *characteristic of node* x_{N+1}.

The results obtained from (6.41) to (6.46) are once again grouped and the corresponding nodal equation (6.38) is written as:

$$\frac{h}{2}\tilde{u}'_{N+1}(t) + \frac{v}{h}\tilde{u}_{N+1}(t) - \frac{v}{h}\tilde{u}_N(t) + \frac{1}{2}\tilde{u}^2_{N+1}(t) - \frac{1}{2}\tilde{u}_N(t)\tilde{u}_{N+1}(t) = vf(t)\,, \tag{6.47}$$

or, after reorganizing the terms, written as follows:

$$\frac{h}{2}\tilde{u}'_{N+1}(t) + \frac{v}{h}[\tilde{u}_{N+1}(t) - \tilde{u}_N(t)] + \frac{1}{2}\tilde{u}_{N+1}(t)[\tilde{u}_{N+1}(t) - \tilde{u}_N(t)] = vf(t)\,, \tag{6.48}$$

d) *Finite differences schemes in space associated with the Neumann condition of the problem* **(CP)**.

Now, discretisation by finite differences of the Neumann condition of the continuous problem **(CP)** defined by (6.1) is performed:

$$\frac{\partial u}{\partial x}(L,t) \equiv \frac{\partial u}{\partial x}(x_{N+1},t) = 0\,, \quad \forall t \geq 0\,. \tag{6.49}$$

To this end, a regressive Taylor's expansion at abscissa x_{N+1}, of the solution u to the problem **(CP)** is written and assumed to be "sufficiently regular" in the neighbourhood of this point.

$$u(x_N,t) = u(x_{N+1},t) - h\frac{\partial u}{\partial x}(x_{N+1},t) + \frac{h^2}{2}\frac{\partial^2 u}{\partial x^2}(x_{N+1},t) + O(h^3)\,. \tag{6.50}$$

This last writing displaying the second partial derivative in x at the point (x_{N+1},t) in order to evaluate the first partial derivative in x at the same point, looks as if a wrong direction has been taken...

Nevertheless, such an expansion was inevitable in order to obtain an approximation by finite differences that is of the same order as the one established for the approximation of the viscous Burgers equation (6.37) i.e. of the second order.

In order to maintain expansion (6.50) while eliminating the second partial derivative of u at point (x_{N+1},t), the Burgers equation is written (that represents a strong hypothesis that should be justified) for this point in order to express the second partial derivative x as follows:

$$\frac{\partial^2 u}{\partial x^2}(x_{N+1},t) = \frac{1}{v}\left[\frac{\partial u}{\partial t}(x_{N+1},t) + u\frac{\partial u}{\partial x}(x_{N+1},t)\right]. \tag{6.51}$$

Taylor's expansion (6.50) is then written as:

$$u(x_N,t) = u(x_{N+1},t) + \ldots,$$
$$-h\frac{\partial u}{\partial x}(x_{N+1},t) + \frac{h^2}{2v}\left[\frac{\partial u}{\partial t}(x_{N+1},t) + u(x_{N+1},t)\frac{\partial u}{\partial x}(x_{N+1},t)\right] + \ldots,$$
$$+O(h^3). \tag{6.52}$$

Thus, the first order partial derivative in x of the solution u at the point (x_{N+1},t) appears at two levels in (6.52).

The first time with a multiple weighting coefficient h, and the second time according to a weighting in h^2.

Insofar an approximation of the second order (after a division by h) is desired, the first order partial derivative is replaced by its value (6.49) and an approximation of the second order is performed for the second first order partial derivative appearing in the equation (6.52).

The first-order approximation of the partial derivative in x at point (x_{N+1},t) is classical and equals:

$$\frac{\partial u}{\partial x}(x_{N+1},t) = \frac{u(x_{N+1},t) - u(x_N,t)}{h} + O(h), \tag{6.53}$$

and from then on, equation (6.52) may be written as:

$$u(x_N,t) = u(x_{N+1},t) - hf(t) + \ldots,$$
$$+\frac{h^2}{2v}\left[\frac{\partial u}{\partial t}(x_{N+1},t) + u(x_{N+1},t)\left(\frac{u(x_{N+1},t) - u(x_N,t)}{h} + O(h)\right)\right] + \ldots,$$
$$+O(h^3). \tag{6.54}$$

Now, the approximations are performed:

$$\tilde{u}_N(t) = \tilde{u}_{N+1}(t) - hf(t) + \frac{h^2}{2v}\left[\tilde{u}'_{N+1}(t) + \tilde{u}_{N+1}(t)\left(\frac{\tilde{u}_{N+1}(t) - \tilde{u}_N(t)}{h}\right)\right]. \tag{6.55}$$

Then, equation (6.55) is rearranged to give it the following form:

$$\frac{h}{2}\tilde{u}'_{N+1}(t) + \frac{v}{h}\left[\tilde{u}_{N+1}(t) - \tilde{u}_N(t)\right] + \frac{1}{2}\tilde{u}_{N+1}(t)\left[\tilde{u}_{N+1}(t) - \tilde{u}_N(t)\right] = vf(t), \tag{6.56}$$

that leads exactly to nodal equation (6.48).

A.7) The differential system $(\mathbf{DS_1}) - (\mathbf{DS_2})$ being of the first order in time and to obtain an approximation by finite differences that is unconditionally stable, it is only necessary to apply a finite differences scheme of the family of θ-scheme for values of θ more than or equal to $1/2$ (cf. D. Euvrard, [4]).

To achieve this, a time step $k \equiv \Delta t$ and the discrete time sequences $t^{(n)}$ are considered, the latter being defined by: $t^{(n)} = kn$.

Moreover, the approximation sequences $\bar{u}_i^{(n)}$ introduced is defined by:

$$\bar{u}_i^{(n)} \simeq \tilde{u}_i(t^{(n)}) \simeq u(x_i, t^{(n)}) \,, \tag{6.57}$$

where the sequences $\tilde{u}_i(t^{(n)})$ is the solution to the differential system $(\mathbf{DS_1}) - (\mathbf{DS_2})$.

Equations (6.31) and (6.48), at time $t^{(n)}$, can formally be written in the form:

$$\tilde{u}_i'(t^{(n)}) = \tilde{\Phi}_i(t^{(n)}), \; \forall i = 1 \text{ to } N+1 \,. \tag{6.58}$$

where function $\tilde{\varphi}$ is defined by:

$$\tilde{\Phi}_i(t^{(n)}) = -\tilde{u}_i(t^{(n)}) \left[\frac{\tilde{u}_{i+1}(t^{(n)}) - \tilde{u}_{i-1}(t^{(n)})}{2h} \right] + \dots ,$$

$$+ v \left[\frac{\tilde{u}_{i-1}(t^{(n)}) - 2\tilde{u}_i(t^{(n)}) + \tilde{u}_{i+1}(t^{(n)})}{h^2} \right] , \; \forall i \neq N+1 \,. \tag{6.59}$$

$$\tilde{\Phi}_{N+1}(t^{(n)}) = \frac{2v}{h} f(t^{(n)}) + \dots$$

$$- \frac{2v}{h^2} \left[\tilde{u}_{N+1}(t^{(n)}) - \tilde{u}_N(t^{(n)}) \right] - \frac{\tilde{u}_{N+1}}{h} \left[\tilde{u}_{N+1}(t^{(n)}) - \tilde{u}_N(t^{(n)}) \right] . \tag{6.60}$$

Then a θ-scheme is applied to functional equation (6.58) in the following way:

$$\bar{u}_i^{(n+1)} = \bar{u}_i^{(n)} + k \left[\theta \bar{\Phi}(t^{(n+1)}) + (1-\theta) \bar{\Phi}(t^{(n)}) \right] , \tag{6.61}$$

where approximation $\bar{\Phi}$ follows the same definition $\tilde{\Phi}$ as that of (6.59)–(6.60), provided that quantities $\tilde{u}_i(t^{(n)})$ are replaced by new approximations $\bar{u}_i^{(n)}$:

$$\bar{\Phi}_i(t^{(n)}) = -\bar{u}_i^{(n)} \left[\frac{\bar{u}_{i+1}^{(n)} - \bar{u}_{i-1}^{(n)}}{2h} \right] + v \left[\frac{\bar{u}_{i-1}^{(n)} - 2\bar{u}_i^{(n)} + \bar{u}_{i+1}^{(n)}}{h^2} \right] , \; \forall i \neq N+1 \,, \tag{6.62}$$

$$\bar{\Phi}_{N+1}(t^{(n)}) = \frac{2v}{h} f(t^{(n)}) - \frac{2v}{h^2} \left[\bar{u}_{N+1}^{(n)} - \bar{u}_N^{(n)} \right] - \frac{\bar{u}_{N+1}}{h} \left[\bar{u}_{N+1}^{(n)} - \bar{u}_N^{(n)} \right] . \tag{6.63}$$

> ▶ *Remark*
>
> Approximation scheme (6.61)–(6.63) is the result of a mixed "finite elements P_1 in space – finite differences in time" approximation.

Given that the mesh in space (6.3) of the interval $[0, L]$ is produced according to a constant step having discretisation h, it has been proved that nodal equations (6.32) and (6.47) associated with each of the basis functions φ_i, $(i = 1$ to $N + 1)$ coincide with a discretisation in space by finite differences.

In other words, in this particular case of a uniform mesh in space, the global approximation system (6.61)–(6.63) is exactly that of a finite differences scheme according to the ordered pair (x, t).

From then on, it is justified to consider the stability of such a numerical scheme according to usual methods applied to evolution equations and solved by finite differences.

But, as presented at the beginning of this question, the choice of a discretisation in time by a θ-scheme is justified to precisely guarantee the stability of the method obtained.

It is then only necessary to consider the values of parameter θ that guarantee the stability scheme (6.61)–(6.63) namely $\theta \geq \dfrac{1}{2}$.

Moreover, it will be noticed that, apart from the particular value $\theta = \dfrac{1}{2}$, the scheme is of the first order in time and of the second order in space.

Finally, when $\theta = \dfrac{1}{2}$, the θ-scheme coincides with that of Crank-Nicholson (cf. D. Euvrard, [4]) and the approximation of system (6.61)–(6.63) is of the second order in time and in space.

6.3 Non-Linear Integro-Differential Equation

6.3.1 Statement

The aim of this problem is to apply the finite elements method in the case of a second order non-linear integro-differential equation.

More precisely, the interest is on the solutions to the following continuous problem (**CP**):

To find $u \in H^2(0,1)$ solution to:

$$(\mathbf{CP}) \begin{cases} -u''(x) + u(x) \displaystyle\int_0^1 u(t)\,dt = f(x)\,, \ 0 \le x \le 1\,, \\ u(0) = 0\,, \ u'(1) = \alpha\,, \end{cases} \qquad (6.64)$$

where f is a given function belonging to $L^2(0,1)$ and α a given parameter.

It is reminded that Sobolev space $H^2(0,1)$ is defined by:

$$H^2(0,1) = \left\{ v \colon [0,1] \to \mathbf{R}\,, v^{(k)} \in L^2(0,1)\,, k = 0 \text{ to } 2 \right\}\,.$$

1) Prove that if u belongs to $H^2(0,1)$, then the integral bearing on u in the continuous problem (**CP**) is convergent.

▶ **Variational Formulation**

2) Let v be a test function defined on $[0,1]$, having real values, belonging to a variational space V. Show that the continuous problem (**CP**) may be written in a variational formulation (**VP**) as:

$$a(u,v) = L(v)\,, \ \forall v \in V\,.$$

non-linear form $a(.,.)$, ***linear*** form $L(.)$ and functional space V has to be specified.

▶ **Lagrange Finite Elements P_1**

3) Approximation of the variational problem (**VP**) is carried out using Lagrange finite elements P_1. To achieve this, a regular mesh of $[0,1]$ interval with a constant step h is introduced, such that:

$$\begin{cases} x_0 = 0,\ x_{N+1} = 1\,, \\ x_{i+1} = x_i + h,\ i = 0 \text{ to } N\,. \end{cases} \qquad (6.65)$$

The approximation space $\tilde{V}$ is now defined using:

$$\tilde{V} = \left\{ \tilde{v} \colon [0,1] \to \mathbf{R},\ \tilde{v} \in C^o([0,1])\,, \tilde{v}|_{[x_i,x_{i+1}]} \in P_1([x_i,x_{i+1}])\,, \tilde{v}(0) = 0 \right\}\,, \quad (6.66)$$

where $P_1([x_i,x_{i+1}])$ refers to the space of polynomials defined over $[x_i,x_{i+1}]$, having a degree less than or equal to one.

– What is the dimension of $\tilde{V}$?

▶ **Approximate Variational Formulation**

4) Let φ_i, ($i = 1$ to dim $\tilde{V}$) be the canonical basis of $\tilde{V}$ verifying $\varphi_i(x_j) = \delta_{ij}$.

After having written the approximate variational formulation $\widetilde{(\mathbf{VP})}$, having solution $\tilde{u}$ and associated with the variational problem $(\mathbf{VP})$, show that by choosing:

$$\tilde{v}(x) = \varphi_i(x)\,, \ (i = 1 \text{ to } \dim \tilde{V}) \quad \text{and} \quad \tilde{u}(x) = \sum_{j=1,\,\dim \tilde{V}} \tilde{u}_j \varphi_j\,, \qquad (6.67)$$

the following $\widetilde{(\mathbf{VP})}$ system is obtained:

$$\widetilde{(\mathbf{VP})} \begin{cases} \displaystyle\sum_{j=1}^{\dim \tilde{V}} A_{ij}\tilde{u}_j + \sum_{j=1}^{\dim \tilde{V}} \sum_{k=1}^{\dim \tilde{V}} B_{ijk}\tilde{u}_j\tilde{u}_k = C_i \,, (\forall i = 1 \text{ to } \dim \tilde{V})\,, & (6.68) \\[2mm] \text{where it was stated:} \\[2mm] \displaystyle A_{ij} = \int_0^1 \varphi_i' \varphi_j' \, \mathrm{d}x\,, \ B_{ijk} = \left(\int_0^1 \varphi_i\varphi_j \, \mathrm{d}x\right) \cdot \left(\int_0^1 \varphi_k \, \mathrm{d}x\right)\,, & (6.69) \\[2mm] \displaystyle C_i = \int_0^1 f\varphi_i \, \mathrm{d}x + \alpha\varphi_i(1)\,. & (6.70) \end{cases}$$

5) Using the trapezium quadrature formula, show that B_{ijk} may be estimated as:

$$B_{ijk} \simeq \begin{vmatrix} hD_{ij}\,, & \forall k = 1 \text{ to } N\,, \\[2mm] \dfrac{h}{2}D_{ij}\,, & \text{if } k = N+1\,, \end{vmatrix} \qquad (6.71)$$

$$\text{where:} \quad D_{ij} = \int_0^1 \varphi_i\varphi_j \, \mathrm{d}x\,.$$

– Infer from this that the approximate variational formulation $\widetilde{(\mathbf{VP})}$ is written as:

$$\widetilde{(\mathbf{VP})} \ \sum_{j=1}^{\dim \tilde{V}} A_{ij}\tilde{u}_j + h\left[\frac{\tilde{u}_{N+1}}{2} + \sum_{k=1}^{N} \tilde{u}_k\right] \cdot \left[\sum_{j=1}^{\dim \tilde{V}} D_{ij}\tilde{u}_j\right] = C_i\,. \qquad (6.72)$$

▶ **Characteristic Basis Function φ_i of a Node Interior at $[0,1]$**

Given the regularity of the mesh, the generic nodal equation of system $\widetilde{(\mathbf{VP})}$ associated with any characteristic basis function φ_i, ($i = 1$ to dim $\tilde{V} - 1$) of a node interior at $[0,1]$ is written as:

$$\widetilde{(\mathbf{VP_{Int}})} \begin{cases} \forall i = 1 \text{ to } \dim \tilde{V} - 1: \\[2mm] A_{i,i-1}\tilde{u}_{i-1} + A_{i,i}\tilde{u}_i + A_{i,i+1}\tilde{u}_{i+1} + \dots \\[2mm] h\left[\dfrac{\tilde{u}_{N+1}}{2} + \displaystyle\sum_{k=1}^{N} \tilde{u}_k\right] \cdot [D_{i,i-1}\tilde{u}_{i-1} + D_{i,i}\tilde{u}_i + D_{i,i+1}\tilde{u}_{i+1}] = C_i\,. \end{cases} \qquad (6.73)$$

– Using the *trapezium formula*, calculate the 7 coefficients (A_{ij}, D_{ij}, C_i).

6) Group the results by writing the corresponding nodal equation.

7) Show that the centred finite differences scheme associated with the differential equation of the continuous problem (**CP**) is obtained again. What is its order of precision?

It is pointed out that the composed trapezium quadrature formula is written as:

$$\int_a^b \xi(s)\,ds \simeq \frac{h}{2}\left[\xi(a)+\xi(b)+2\sum_{i=1}^N \xi(x_i)\right].$$

▶ **Characteristic Basis Function φ_{N+1} of the Node at Abscissa x_{N+1}**

8) The same procedure is followed for the basis function φ_{N+1} characterising the final node x_{N+1}.

The corresponding equation of system $\widetilde{(\mathbf{VP})}$ is then written as:

$$\widetilde{(\mathbf{VP})}_{N+1} \begin{cases} A_{N+1,N}\tilde{u}_N + A_{N+1,N+1}\tilde{u}_{N+1} + \dots \\[2mm] h\left[\dfrac{\tilde{u}_{N+1}}{2} + \displaystyle\sum_{k=1}^N \tilde{u}_k\right]\cdot\left[D_{N+1,N}\tilde{u}_N + D_{N+1,N+1}\tilde{u}_{N+1}\right] = C_{N+1}\,. \end{cases} \tag{6.74}$$

– Using **the trapezium formula**, calculate the 5 coefficients $A_{N+1,N}$, $A_{N+1,N+1}$, $D_{N+1,N}$, $D_{N+1,N+1}$ and C_{N+1}.

9) Group the results by writing the corresponding nodal equation.

10) Using the finite differences method, find again this nodal equation by carrying a second order discretisation of the Neumann boundary conditions of continuous problem (**CP**) when $x_{N+1} = 1$.

6.3.2 Solution

A.1) The integro-differential equation of continuous problem **(CP)** presents a non-linearity inherent to the matching term between u and the integral $\int_0^1 u(x)\,dx$.

It is then noticed that the convergence of this integral is provided by the functional space in which the continuous problem is set, i. e. $H^2(0,1)$.

In fact, using the Cauchy-Schwartz inequality, the following is obtained:

$$\left| \int_0^1 1 \cdot u(x)\,dx \right| \leq \left[\int_0^1 |1|^2\,dx \right]^{1/2} \cdot \left[\int_0^1 |u(x)|^2\,dx \right]^{1/2} \leq \left[\int_0^1 |u(x)|^2\,dx \right]^{1/2}. \quad (6.75)$$

In other words, if solution u of continuous problem **(CP)** is searched for in the Sobolev space $H^2(0,1)$, u *de facto* belongs to $L^2(0,1)$ and subsequently to $L^1(0,1)$ according to inequality (6.75).

A.2) Let v be a test function defined on $[0,1]$ and having real values. The integro-differential equation of continuous problem **(CP)** is multiplied by v and the obtained equation is integrated between 0 and 1.

$$-\int_0^1 u''v\,dx + \int_0^1 \left(\int_0^1 u(s)\,ds \right) uv\,dx = \int_0^1 fv\,dx. \quad (6.76)$$

As usual, once the variational formulation is definitely established, the functional space V will be specified.

Moreover, an integration by parts using the Neumann condition $u'(1) = \alpha$ enables the writing of:

$$\int_0^1 \left[u'v' + \left(\int_0^1 u(s)\,ds \right) uv \right] dx + u'(0)v(0) = \int_0^1 fv\,dx + \alpha v(1). \quad (6.77)$$

Given that this formulation does not allow the homogenous Dirichlet problem $u(0) = 0$ to be taken into account, functions v are made to satisfy: $v(0) = 0$.

The variational problem **(VP)** is thus written as:

$$\textbf{(VP)} \begin{cases} \text{Find } u \text{ belonging to } V \text{ solution of: } a(u,v) = L(v),\ \forall v \in V,\ \text{where:} \\[2mm] a(u,v) \equiv \int_0^1 \left[u'(x)v'(x) + \left(\int_0^1 u(s)\,ds \right) u(x)v(x) \right] dx, \\[2mm] L(v) \equiv \int_0^1 f(x)v(x)\,dx + \alpha v(1). \end{cases}$$

$$(6.78)$$

The functional framework V in which a variational formulation **(VP)** makes sense is now defined.

Concerning the integrals bearing on $u'v'$, on one hand, and on fv, on the other hand, it was often observed that (cf. Dirichlet [3.1] or Neumann [3.2] problems), the Cauchy-Schwartz inequality enabled the existence of these two integrals to be guaranteed.

As for the integral bearing on the non-linear term, it is only necessary to note that:

$$\int_0^1 \left[\left(\int_0^1 u(s)\,ds \right) u(x)v(x) \right] dx = \left(\int_0^1 u(s)\,ds \right) \cdot \left(\int_0^1 u(x)v(x)\,dx \right) . \qquad (6.79)$$

Once again, the convergence of the integral is then ensured by the Cauchy-Schwartz inequality applied to the integral bearing on uv.

Thus, functional space V that allows giving a sense to variational formulation **(VP)** is defined by:

$$V \equiv H^1(0,1) \cap \{v \colon [0,1] \to \mathbf{R} \,, \ v(0) = 0\} . \qquad (6.80)$$

A.3) The dimension of approximation space $\tilde{V}$ may be found out through several means. The simplest way consists in noting that functions $\tilde{v}$ of $\tilde{V}$ are essentially pecked lines, in fact, affine per entire mesh $[x_i, x_{i+1}]$ and are zero when $x = 0$.

Therefore, having $(N+2)$ discretisation points for the whole mesh of the $[0,1]$ interval, two functions of $\tilde{V}$ distinguish themselves by the difference in their values that may be observed at $(N+1)$ points $(x_1, \ldots, x_{N+1})$ and, in addition, any function $\tilde{v}$ of V should satisfy $\tilde{v}_0 = 0$.

In other words, a function $\tilde{v}$ belonging to $\tilde{V}$ is completely determined by the $(N+1)$-tuple $(\tilde{v}_1, \ldots, \tilde{v}_{N+1})$.

This implies that space is $\tilde{V}$ isomorphic at $\mathbf{R}^{N+1}$. In conclusion, it is inferred from this that the dimension of $\tilde{V}$ is equal to $(N+1)$.

A.4) The approximate variational formulation $\widetilde{(\mathbf{VP})}$ is obtained by substituting approximate functions $\tilde{u}$ and $\tilde{v}$ by functions u and v in the variational formulation **(VP)**.

Moreover, the expressions given by (6.67) are used and the following is obtained:

$$\widetilde{(\mathbf{VP})} \quad \left[\begin{array}{l} \text{Find } (\tilde{u}_j)(j = 1 \text{ to } N+1) \,, \text{ solution to:} \\[2mm] \displaystyle\sum_{j=1,N+1} \left[\int_0^1 \left\{ \varphi_j'\varphi_i'(x) + \left(\sum_{k=1,N+1} \int_0^1 \tilde{u}_k\varphi_k(s)\,ds \right) \varphi_j(x)\varphi_i(x) \right\} dx \right] \tilde{u}_j \\[4mm] = \displaystyle\int_0^1 f(x)\varphi_i(x)\,dx + \alpha\varphi_i(1) . \end{array} \right.$$

$$(6.81)$$

It is then only necessary to observe that the integral bearing on $\tilde{u}_k\varphi_k(s)$ is independent of the variable x of the main integral in expression (6.81).

By identification, expressions (6.69) and (6.70) are then obtained.

A.5) In order to estimate the quantities B_{ijk} by approximation, the trapezium quadrature formula is used to approximate the integral bearing on function φ_k:

$$\left[\begin{array}{l} \forall k = 1 \text{ to } N: \\ \displaystyle\int_0^1 \varphi_k(s)\,ds = \int_{x_{k-1}}^{x_{k+1}} \varphi_k(s)\,ds = \int_{x_{k-1}}^{x_k} \varphi_k(s)\,ds + \int_{x_k}^{x_{k+1}} \varphi_k(s)\,ds, \\ \qquad \simeq \dfrac{h}{2}[1+0] + \dfrac{h}{2}[0+1] = h. \end{array} \right. \qquad (6.82)$$

The case of basis function φ_{N+1} should be treated separately because its support is solely constituted by the $[x_N, x_{N+1}]$ interval (cf. Fig. 6.3).

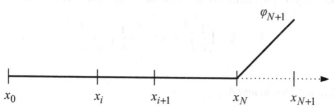

Fig. 6.3 Basis Function φ_{N+1}

The following is then inferred:

$$\int_0^1 \varphi_{N+1}(s)\,ds = \int_{x_N}^{x_{N+1}} \varphi_{N+1}(s)\,ds \simeq \frac{h}{2}[1+0] = \frac{h}{2}. \qquad (6.83)$$

The evaluations of equations (6.82)–(6.83) are then injected in the generic equation of the approximate variational problem (6.81) and equation (6.72) is obtained.

A.6) The basis functions $\varphi_i, (i = 1 \text{ to } N)$ characterising the nodes of the mesh strictly interior at the $[0, 1]$ integration interval are now considered.

The generic equation of system (6.81) contains terms that are, *a priori*, non-zero and that only correspond to functions φ_j whose support intercepts that of a given function φ_i, (cf. Fig. 6.4).

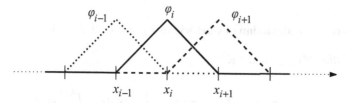

Fig. 6.4 Basis Functions φ_{i-1}, φ_i and φ_{i+1}

In other words, basis functions φ_{i-1}, φ_i and φ_{i+1} are those concerned.

That is why equation $(\widetilde{\text{VP}}_{\text{Int}})$ contains only terms $A_{i,i-1}$, $A_{i,i}$ and $A_{i,i+1}$ on one hand and $D_{i,i-1}$, $D_{i,i}$ and $D_{i,i+1}$, on the other hand.

▶ **Exact calculation of coefficients** $A_{ij}, j = i-1, i, i+1$

a) *Calculation of coefficient* A_{ii}.

$$A_{ii} = \int_0^1 (\varphi_i')^2 \, dx = \int_{\text{Supp } \varphi_i'} (\varphi_i')^2 \, dx = \int_{x_{i-1}}^{x_i} (\varphi_i')^2 \, dx + \int_{x_i}^{x_{i+1}} (\varphi_i')^2 \, dx . \quad (6.84)$$

Since basis functions φ_i of $\tilde{V}$ are piecewise affine, derivatives φ_i are constant on every mesh having the form $[x_i, x_{i+1}]$.

Then, what may be done is either evaluating every integral of equation (6.84) or applying the trapezium quadrature formula that is exact for the constant functions:

$$A_{ii} = h \times \left(\frac{1}{h}\right)^2 + h \times \left(-\frac{1}{h}\right)^2 = \frac{2}{h} . \quad (6.85)$$

b) *Calculation of coefficient* $A_{i,i-1}$.

$$A_{i,i-1} = \int_0^1 \varphi_i' \varphi_{i-1}' \, dx = \int_{\text{Supp } \varphi_i \cap \text{Supp } \varphi_{i-1}'} \varphi_i' \varphi_{i-1}' \, dx = \int_{x_{i-1}}^{x_i} \varphi_i' \varphi_{i-1}' \, dx$$

$$= h \times \left(\frac{1}{h}\right) \times \left(-\frac{1}{h}\right) = -\frac{1}{h} . \quad (6.86)$$

c) *Calculation of coefficient* $A_{i,i+1}$.

$A_{i,i+1}$ is obtained directly because it is only necessary to note that:

$$A_{i,i+1} = A_{i+1,i} = A_{i,i-1} . \quad (6.87)$$

To achieve this, the symmetry of Matrix $A_{i,j}$ as well as the invariance by horizontal translation along the mesh would have been used as a result of the uniformity of its discretisation.

▶ **Approximate Calculation of Coefficients** $D_{ij}, j = i-1, i, i+1$

a) *Calculation of coefficient* D_{ii}.

$$D_{ii} = \int_0^1 \varphi_i^2 \, dx = \int_{\text{Supp } \varphi_i} \varphi_i^2 \, dx = \int_{x_{i-1}}^{x_i} \varphi_i^2 \, dx + \int_{x_i}^{x_{i+1}} \varphi_i^2 \, dx$$

$$\simeq \frac{h}{2}(0+1) + \frac{h}{2}(1+0) = h . \quad (6.88)$$

b) *Calculation of coefficient* $D_{i,i-1}$.

$$D_{i,i-1} = \int_0^1 \varphi_i \varphi_{i-1} \, dx = \int_{\text{Supp } \varphi_i \, \cap \, \text{Supp } \varphi_{i-1}} \varphi_i \varphi_{i-1} \, dx = \int_{x_{i-1}}^{x_i} \varphi_i \varphi_{i-1} \, dx$$

$$\simeq \frac{h}{2}(0 \times 1 + 1 \times 0) = 0 \, . \tag{6.89}$$

c) *Calculation of coefficient* $D_{i,i+1}$.

Because of symmetry reasons similar to those brought up for the calculation of coefficients $A_{i,i+1}$, the following is obtained:

$$D_{i,i+1} = D_{i+1,i} = D_{i,i-1} \equiv 0 \, . \tag{6.90}$$

▶ **Calculation of the second member** C_i

Given the following basis functions property: $\forall i = 1$ to N: $\varphi_i(1) = 0$, the second member C_i is estimated in the following way:

$$C_i = \int_0^1 f \varphi_i \, dx = \int_{x_{i-1}}^{x_i} f \varphi_i \, dx + \int_{x_i}^{x_{i+1}} f \varphi_i \, dx$$

$$\simeq \frac{h}{2}[0 + f_i] + \frac{h}{2}[f_i + 0]$$

$$C_i \simeq h f_i \, . \tag{6.91}$$

A.7) The nodal equation associated with a basis function φ_i, $(i = 1$ to $N)$ is obtained by grouping all the results of the previous question:

$$\forall i = 1, N: \; -\left[\frac{\tilde{u}_{i-1} - 2\tilde{u}_i + \tilde{u}_{i+1}}{h^2}\right] + h \left[\frac{\tilde{u}_{N+1}}{2} + \sum_{k=1}^N \tilde{u}_k\right] \cdot \tilde{u}_i = f_i \, . \tag{6.92}$$

A.8) Now, it is proposed to find again nodal equation (6.92) associated with any basis function φ_i, $(i = 1$ to $N)$ by applying the finite différences method.

To achieve this, it is advisable to write the integro-differential equation of continuous problem **(CP)** at point x_i, then to proceed to the approximation of the second derivative of u on one hand and the approximation of the integral of solution u on the interval $[0, 1]$ on the other hand.

Concerning the second derivative, Taylor's formula is simultaneously used progressively and regressively and this gives:

$$u(x_{i+1}) = u(x_i) + h u'(x_i) + \frac{h^2}{2} u''(x_i) + \frac{h^3}{3!} u^3(x_i) + O(h^4) \, , \tag{6.93}$$

$$u(x_{i-1}) = u(x_i) - h u'(x_i) + \frac{h^2}{2} u''(x_i) - \frac{h^3}{3!} u^3(x_i) + O(h^4) \, . \tag{6.94}$$

Then, the sum between (6.93) and (6.94) gives:

$$u''(x_i) = \frac{u(x_{i-1}) - 2u(x_i) + u(x_{i+1})}{h^2} + O(h^2) . \qquad (6.95)$$

Concerning the integral of u between 0 and 1, the composed trapezium quadrature formula is used, by noting that solution u verifies the homogenous Dirichlet condition when $x = 0$:

$$\int_0^1 u(s)\, ds = \frac{h}{2}\left[u(x_{N+1}) + 2\sum_{i=1}^{N} u(x_i) \right] + O(h^2) . \qquad (6.96)$$

Then, the integro-differential equation of continuous problem **(CP)** at point x_i is written, $u''(x_i)$ by substituting by (6.95) and $\int_0^1 u(s)\, ds$ by (6.96) to give:

$$-\left[\frac{u(x_{i-1}) - 2u(x_i) + u(x_{i+1})}{h^2} \right] + u(x_i) \cdot \frac{h}{2}\left[u(x_{N+1}) + 2\sum_{i=1}^{N} u(x_i) \right]$$

$$= f(x_i) + O(h^2) . \qquad (6.97)$$

The next step consists in writing the finite differences scheme by substituting the approximation series $\tilde{u}_i$ by the real values $u(x_i)$.

This operation enables one to eliminate the residue of the second order $O(h^2)$ in equation (6.97) and exactly yields nodal equation (6.92) which was found by the finite elements method.

Moreover, finite differences scheme (6.92) is of the second order as a result of the elimination of the term in $O(h^2)$.

▶ **A general observation.**

As usual, for the use of Taylor's formula in such a context, it will be noticed that more regularity was assumed for solution u of continuous problem **(CP)**, (assuming in the present case that u is at least C^4 over $]0, 1[$ so that the writing of Taylor's formula up to fourth-order would be possible), although it seems that, initially at least, the latter is rather C^2 over $]0, 1[$.

In fact, although it may be possible in some cases to establish that solution u possesses more regularity than it seems to have, the majority of cases need to be explained because the regularity of solution u depends on one hand on the regularity of the second member and on the regularity inherent to the structure of the differential operator on the other hand.

How then can the free choice of the appropriate regularity for solution u of the continuous problem be explained in order to write a Taylor's formula, that may concern a majority of differential equations which do not permit such a choice?

The solution is not totally mind-satisfying, however, mastery over the compromise is often what is needed in matters concerning numerical analysis (like in many other fields, be it scientific or not!): The Taylor's expansions that have been written are valid only for the category of differential equations presenting "sufficiently regular" solutions.

To such a point that when a differential equation presents a solution whose regularity may be proved to be different from the one required by Taylor's formula, such a writing is knowingly maintained given that what is ultimately necessary is to build a sequence of probable approximations that tends to the exact solution of the problem.

Thus, having assumed that there is certainly as much regularity as possible, an approximation of the differential operator may be then proposed by an algebraic procedure that would provide an approximation of a certain standard.
This standard known as the order of the finite differences scheme in the jargon of numerical analysis finally enables the measurement and appreciation of the performances of various finite differences schemes having equal data and sufficiency of regularity of solutions to differential equations.

▶ **Characteristic Basis Function φ_{N+1} of the Node at Abscissa x_{N+1}**

A.9) The equation of system $\widetilde{(\text{VP})}$ defined by (6.72) for the particular basis function φ_{N+1}, is now written and leads to equation (6.74).

The 5 coefficients of this nodal equation are evaluated in the same way as for the calculations presented in the previous question.

▶ **Exact Calculation of Coefficients $A_{N+1,N+1}$ and $A_{N+1,N}$**

a) *Calculation of coefficient $A_{N+1,N+1}$.*

$$A_{N+1,N+1} = \int_0^1 (\varphi'_{N+1})^2 \, dx = \int_{\text{Supp } \varphi'_{N+1}} (\varphi'_{N+1})^2 \, dx = \int_{x_N}^{x_{N+1}} (\varphi'_{N+1})^2 \, dx$$

$$= h \times \left(\frac{1}{h}\right)^2 = \frac{1}{h}. \tag{6.98}$$

b) *Calculation of coefficient $A_{N+1,N}$.*

$$A_{N+1,N} = \int_0^1 \varphi'_{N+1} \varphi'_N \, dx = \int_{\text{Supp } \varphi'_{N+1} \cap \text{Supp } \varphi'_N} \varphi'_{N+1} \varphi'_N \, dx$$

$$= \int_{x_N}^{x_{N+1}} \varphi'_{N+1} \varphi'_N \, dx = h \times \left(\frac{1}{h}\right) \times \left(-\frac{1}{h}\right) = -\frac{1}{h}. \tag{6.99}$$

▶ **Approximate Calculation of Coefficients $D_{N+1,N+1}$ and $D_{N+1,N}$**

a) *Calculation of coefficient $D_{N+1,N+1}$.*

$$D_{N+1,N+1} = \int_0^1 \varphi_{N+1}^2 \, dx = \int_{\text{Supp } \varphi_{N+1}} \varphi_{N+1}^2 \, dx = \int_{x_N}^{x_{N+1}} \varphi_{N+1}^2 \, dx$$

$$\simeq \frac{h}{2}(0+1) = \frac{h}{2} \, . \tag{6.100}$$

b) *Calculation of coefficient $D_{N+1,N}$.*

$$D_{N+1,N} = \int_0^1 \varphi_{N+1} \varphi_N \, dx = \int_{\text{Supp } \varphi_{N+1} \cap \text{Supp } \varphi_N} \varphi_{N+1} \varphi_N \, dx$$

$$= \int_{x_N}^{x_{N+1}} \varphi_{N+1} \varphi_N \, dx \simeq \frac{h}{2}(0 \times 1 + 1 \times 0) = 0 \, . \tag{6.101}$$

▶ **Calculation of the Second Member C_{N+1}**

$$C_{N+1} = \int_0^1 f \varphi_{N+1} \, dx = \int_{x_N}^{x_{N+1}} f \varphi_{N+1} \, dx + \alpha \, ,$$

$$\simeq \frac{h}{2}[0 + f_{N+1}] + \alpha \, ,$$

$$C_{N+1} \simeq \frac{h}{2} f_{N+1} + \alpha \, . \tag{6.102}$$

A.10) Then, all the results of the previous question are grouped and the nodal equation corresponding to the basis function φ_{N+1} is built up:

$$-\frac{1}{h}\tilde{u}_N + \frac{1}{h}\tilde{u}_{N+1} + h\left[\frac{\tilde{u}_{N+1}}{2} + \sum_{k=1}^N \tilde{u}_k\right] \cdot \frac{h}{2}\tilde{u}_{N+1} = \frac{h}{2}f_{N+1} + \alpha \, . \tag{6.103}$$

By rearranging the terms of (6.103), the latter may be written in the form:

$$\frac{2}{h^3}[\tilde{u}_{N+1} - \tilde{u}_N] + \frac{\tilde{u}_{N+1}^2}{2} + \tilde{u}_{N+1}\sum_{k=1}^N \tilde{u}_k = \frac{f_{N+1}}{h} + \frac{2\alpha}{h^2} \, . \tag{6.104}$$

A.11) The discretisation of the Neumann condition $u'(1) = \alpha$ using finite differences is now carried out.

To achieve this, a third-order regressive Taylor's expansion is performed in order to globally maintain a second-order method for the approximation of the continuous problem **(CP)** using the finite differences method.

$$u(x_N) = u(x_{N+1}) - hu'(x_{N+1}) + \frac{h^2}{2}u''(x_{N+1}) + O(h^3) \, . \tag{6.105}$$

Moreover, it is assumed that the integro-differential equation may be written in $x = 1$ and once again, this presupposes properties of regularity of solution u of the continuous problem (**CP**).

Thus, the second derivative u at abscissa x_{N+1} may be replaced according to the values of u taken at the other nodes of the mesh.

$$u(x_N) = u(x_{N+1}) - \alpha h + \frac{h^2}{2}\left[u(x_{N+1}) \int_0^1 u(s)\, ds - f(x_{N+1}) \right] + O(h^3), \quad (6.106)$$

where it has been set that:

$$f_{N+1} = f(x_{N+1}).$$

Then, after the use of composed trapezium quadrature formula, finally gives:

$$u(x_N) = u(x_{N+1}) - \alpha h - \frac{h^2}{2} f(x_{N+1}) + \frac{h^3}{2} u(x_{N+1}) \left[\frac{u(x_{N+1})}{2} + \sum_{k=1}^{N} \tilde{u}_k \right] + O(h^3).$$

$$(6.107)$$

Finally, equation (6.107) is multiplied by $2/h^3$ and its various terms are rearranged in order to exactly find nodal equation (6.104).

6.4 Riccati Differential Equation

6.4.1 Statement

This problem is dedicated to the numerical resolution of the Riccati non-linear differential equation using the method of finite elements P_1.

In other words, the interest is on the scalar function u of variable x, being solution to the continuous problem **(CP)**:

Find $u \in H^1(0,1)$ as solution to:

$$\textbf{(CP)} \begin{cases} u'(x) + u^2(x) = f(x) \ 0 \le x \le 1\,, \\ u(0) = 0\,, \end{cases} \tag{6.108}$$

where f is a given function belonging to $L^2(0,1)$.

1) Let v be a test function defined by $[0,1]$, having real values and belonging to variational space V. Show that the continuous problem **(CP)** may be written in a variational formulation **(VP)** to be given later.

– What are the properties to be verified by functions v belonging to V?

▶ **Lagrange Finite Elements P_1**

2) Approximation of variational problem **(VP)** is carried out using Lagrange finite elements P_1. To achieve this, a regular mesh of the $[0,1]$ interval with a constant step h is introduced, such that:

$$\begin{cases} x_0 = 0\,, \ x_{N+1} = 1\,, \\ x_{i+1} = x_i + h\,, \ i = 0 \text{ to } N\,. \end{cases} \tag{6.109}$$

The approximation space $\tilde{V}$ is now defined using:

$$\tilde{V} = \{\tilde{v}\colon [0,1] \to \mathbf{R}\,, \ \tilde{v} \in C^0([0,1])\,, \ \tilde{v}|_{[x_i,x_{i+1}]} \in P_1([x_i,x_{i+1}])\,, \ \tilde{v}(0) = 0\}\,, \tag{6.110}$$

where $P_1([x_i,x_{i+1}])$ refers to the space of polynomials defined on $[x_i,x_{i+1}]$, having a degree less than or equal to one.

– What is the dimension of $\tilde{V}$?

3) Let $\varphi_i, (i = 1$ to $\dim \tilde{V})$ be the basis of $\tilde{V}$ verifying $\varphi_i(x_j) = \delta_{ij}$.

After writing the approximate variational formulation of solution $\tilde{u}$ associated with problem **(VP)**, show that by choosing:

$$\tilde{v}(x) = \varphi_i(x) \quad \text{and} \quad \tilde{u}(x) = \sum_{j=1,\,\dim \tilde{V}} \tilde{u}_j \varphi_j\,, \tag{6.111}$$

the following $\widetilde{(\mathbf{VP})}$ system is obtained:

$$\widetilde{(\mathbf{VP})} \begin{bmatrix} \sum_{j=1,\,\dim \tilde{V}} A_{ij}\tilde{u}_j + \sum_{(j,k)\in\{1,\dots\dim \tilde{V}\}} B_{ijk}\tilde{u}_j\tilde{u}_k = C_i\,, & (6.112) \\[2mm] \forall i \in \{1,\dots,\tilde{V}\}\,,\text{ where it was stated:} \\[2mm] A_{ij} = \int_0^1 \varphi_i\varphi_j'\,dx\,,\ B_{ijk} = \int_0^1 \varphi_i\varphi_j\varphi_k\,dx \\[2mm] C_i = \int_0^1 f\varphi_i\,dx\,. & (6.113) \end{bmatrix}$$

– What is the characteristic of the $\widetilde{(\mathbf{VP})}$ system?

▶ **Characteristic Basis Function φ_i of a Node Strictly Interior at [0,1]**

4) Given the regularity of the mesh, the generic nodal equation of system $\widetilde{(\mathbf{VP})}$ associated with any characteristic basis function φ_i of a node strictly interior at $[0,1]$ is written as $\widetilde{(\mathbf{VP_{Int}})}$:

$$\begin{cases} \forall i = 1 \text{ to } \dim \tilde{V} - 1: \\[2mm] A_{i,i-1}\,\tilde{u}_{i-1} + A_{i,i}\,\tilde{u}_i + A_{i,i+1}\,\tilde{u}_{i+1} + \dots \\[2mm] (B_{i,i-1,i} + B_{i,i,i-1})\,\tilde{u}_i\tilde{u}_{i-1} + B_{i,i-1,i-1}\,\tilde{u}_{i-1}^2 + B_{i,i,i}\,\tilde{u}_i^2 + \dots \\[2mm] (B_{i,i,i+1} + B_{i,i+1,i})\,\tilde{u}_i\tilde{u}_{i+1} + B_{i,i+1,i+1}\,\tilde{u}_{i+1}^2 = C_i\,. \end{cases} \qquad (6.114)$$

– Using the trapezium formula, calculate the 11 coefficients $(A_{i,j},\ B_{i,j,k},\ C_i)$.

5) Group the results by writing the corresponding nodal equation.

6) Show that the centred finite differences scheme associated with the Riccati differential equation of problem **(CP)** is obtained again. What is its order of precision?

Remember that the trapezium quadrature formula is written as:

$$\int_a^b \xi(s)\,ds \simeq \frac{(b-a)}{2}\{\xi(a) + \xi(b)\}\,.$$

▶ **Characteristic Basis Function φ_{N+1} of the Node at Abscissa x_{N+1}**

7) The same procedure is followed for the basis function φ_{N+1} characterising the final node x_{N+1}.

The corresponding equation of system $\widetilde{(\mathbf{VP})}$ is then written as:

$$\widetilde{(\mathbf{VP_{Ext}})} \begin{cases} A_{N+1,N}\,\tilde{u}_N + A_{N+1,N+1}\,\tilde{u}_{N+1} + B_{N+1,N,N}\,\tilde{u}_N^2 + \dots \\[2mm] (B_{N+1,N,N+1} + B_{N+1,N+1,N})\,\tilde{u}_N\tilde{u}_{N+1} + B_{N+1,N+1,N+1}\,\tilde{u}_{N+1}^2 = C_{N+1}\,. \end{cases}$$
$$(6.115)$$

– Using the trapezium formula, calculate the 7 coefficients $(A_{i,j}, B_{i,j,k}, C_i)$ of equation $(\widetilde{VP_{Ext}})$.

8) Group the results by writing the corresponding nodal equation.

9) Show that the nodal equation $(\widetilde{VP_{Ext}})$ may be obtained again by using the finite differences method. What is the order of approximation of the scheme thus obtained?

6.4.2 Solution

▶ **Theoretical Part**

A.1) Let v be a defined function of the $[0, 1]$ interval, having real values and belonging to a variational space V. The characterization of V will be worked out, *a posteriori*, once the variational formulation is formally established.

The Riccati differential equation of continuous problem **(CP)** (6.108) is multiplied by v and the equation thus obtained is integrated between 0 and 1.

A variational formulation **(VP)** may thus be written as:

Find $u \in V$ solution to:

$$\int_0^1 u'(x)v(x)\,dx + \int_0^1 u^2(x)v(x)\,dx = \int_0^1 f(x) \cdot v(x)\,dx, \quad \forall v \in V. \tag{6.116}$$

Before specifying the nature of space V, note that given the structure of the two integrands $u'v$ and u^2v, no integration by parts would lead to the achievement of a more exploitable variational formulation.

That is why variational equation (6.116) will be maintained throughout the rest of this exercise.

The regularity of functions v of V is now dealt with in order to establish sufficient conditions for the existence of the integrals of equation (6.116).

The first integral of equation (6.116) is controlled by applying the Cauchy-Schwartz inequality as below:

$$\left| \int_0^1 u'(x)v(x)\,dx \right| \le \int_0^1 |u'(x)v(x)|\,dx$$

$$\le \left[\int_0^1 |u'(x)|^2\,dx \right]^{1/2} \cdot \left[\int_0^1 |v(x)|^2\,dx \right]^{1/2}. \tag{6.117}$$

Thus, should u' and v be two functions for which the regularity of the functions of $L^2(0, 1)$ is set, then integral $\int_0^1 u'(x)v(x)\,dx$ is convergent.

As for the second integral, the procedure is as follows:

$$\left| \int_0^1 u^2(x)v(x)\,dx \right| \le \int_0^1 |u^2(x)v(x)|\,dx \le C \int_0^1 |v(x)|\,dx$$

$$\le C \left[\int_0^1 |v(x)|^2\,dx \right]^{1/2}, \tag{6.118}$$

where it would be stated:

$$C = \max_{x \in [0,1]} \left[u^2(x) \right].$$

In other words, considering u and v as describing space $H^1(0,1)$, on one hand, will lead to the continuous Sobolev injection:

$$H^1(0,1) \subseteq C^0([0,1]) , \tag{6.119}$$

which enables the justification of the introduction of finite constant C as the maximum of function u^2.

On the other hand, integral $\int_0^1 u'(x)v(x)\,dx$ will consequently be convergent.

Finally, in order to maintain the whole information existing in the formulation of continuous problem **(CP)** i. e. the Dirichlet boundary conditions $u(0) = 0$, functions v of V are made to satisfy the same boundary conditions $(v(0) = 0)$ so that solution u of variational problem **(VP)** maintains this property of homogeneity when $x = 0$ as one of the particular functions v of V.

The following is thus stated:

$$V \equiv H_0^1(0,1) = H^1(0,1) \cap \{v/v(0) = 0\} .$$

▶ **Lagrange Finite Elements P$_1$**

A.2) There are several ways to find the dimension of approximation space $\tilde{V}$. The simplest way consists in noting that functions $\tilde{v}$ of $\tilde{V}$ are essentially pecked lines, in fact, affine per entire mesh $[x_i, x_{i+1}]$ and are zero when $x = 0$.

Therefore, having $(N + 2)$ discretisation points for the whole mesh of the interval $[0,1]$, two functions of $\tilde{V}$ distinguish themselves by the difference in their values that may be observed at $(N + 1)$ points $(x_1, \ldots, x_{N+1})$ and in addition, any function $\tilde{v}$ of V should satisfy $\tilde{v}_0 = 0$.

In other words, function $\tilde{v}$ belonging to $\tilde{V}$ is entirely determined by $(N + 1)$-tuple $(\tilde{v}_1, \ldots, \tilde{v}_{N+1})$, the trace of its values at the nodes of discretisation $(x_1, \ldots, x_{N+1})$.

This implies that space $\tilde{V}$ is isomorphic at $\mathbf{R}^{N+1}$. In conclusion, it is inferred from this that the dimension of $\tilde{V}$ is equal to $(N + 1)$.

A.3) The approximate variational formulation $\widetilde{(\text{VP})}$ results from the substitution of the pair of functions (u,v) belonging to $V \times V$ in the variational formulation **(VP)** by functions $(\tilde{u}, \tilde{v})$ describing $\tilde{V} \times \tilde{V}$:

$$\int_0^1 \tilde{u}'(x)\tilde{v}(x)\,dx + \int_0^1 \tilde{u}^2(x)\tilde{v}(x)\,dx = \int_0^1 f(x) \cdot \tilde{v}(x)\,dx , \quad \forall \tilde{v} \in \tilde{V} . \tag{6.120}$$

$\tilde{v} = \varphi_i$ is now chosen and $\tilde{u}$ broken down on the basis functions φ_j of $\tilde{V}$ to immediately lead to the variational formulation $\widetilde{(\text{VP})}$ defined by (6.112)–(6.113).

The major characteristic of algebraic system (6.112)–(6.113) having unknowns $(\tilde{u}_1, \ldots, \tilde{u}_{N+1})$ is its non-linearity with regard to these unknowns.

Of course, traces of the non-linearity of Riccati differential equation are found after discretisation.

This non-linearity does pause a problem, because resolution methods of the Gauss, Jacobi, Gauss-Seidel types, successive relaxation methods or simple or conjugated gradient methods may not be directly operational in such a case.

Then the Newton type of methods should be resorted to, (e. g. the work of M. Crouzeix and A. L. Mignot, [2] may be consulted).

▶ **Characteristic Basis Function φ_i of a Node Strictly Interior at [0,1]**

A.4) Equations corresponding to the basis functions φ_i characterising a node (i. e. equal to one at the concerned node and zero at the other nodes) strictly interior at the mesh are now selected from the systems (6.112)–(6.113).

In other words, in the present case, the values of i having values between 1 and N are concerned.

Moreover, given that the support of a function φ_i only contains segment $[x_{i-1}, x_{i+1}]$, only the basis functions φ_{i-1}, φ_i and φ_{i+1} may produce non-zero contributions in the calculation of coefficients A_{ij} and B_{ijk}.

In fact, the following is obtained:

$$A_{ij} = \int_0^1 \varphi_i \varphi_j' \, dx = \int_{\text{Supp } \varphi_i \,\cap\, \text{Supp } \varphi_j'} \varphi_i \varphi_j' \, dx = \int_{\text{Supp } \varphi_i \,\cap\, \text{Supp } \varphi_j} \varphi_i \varphi_j' \, dx$$

$$B_{ijk} = \int_0^1 \varphi_i \varphi_j \varphi_k \, dx = \int_{\text{Supp } \varphi_i \,\cap\, \text{Supp } \varphi_j \,\cap\, \text{Supp } \varphi_k} \varphi_i \varphi_j \varphi_k \, dx. \tag{6.121}$$

Thus, when i is fixed, (i ranging from 1 to N, since it concerns only interior nodes), the corresponding equation of system (6.112-6.113) is written according to formula (6.114).

Coefficients A_{ij} and B_{ijk} are now calculated using the trapezium rule for the approximation.

▶ **Approximate Calculation of Coefficients A_{ij}, $(j = i - 1, i, i + 1)$**

a) *Calculation of coefficient A_{ii}.*

$$A_{ii} = \int_0^1 \varphi_i \varphi_i' \, dx = \int_{\text{Supp } \varphi_i} \varphi_i \varphi_i' \, dx = \int_{x_{i-1}}^{x_i} \varphi_i \varphi_i' \, dx + \int_{x_i}^{x_{i+1}} \varphi_i \varphi_i' \, dx$$

$$A_{ii} \simeq \frac{1}{h} \times \frac{h}{2} \{\varphi_i(x_{i-1}) + \varphi_i(x_i)\} - \frac{1}{h} \times \frac{h}{2} \{\varphi_i(x_i) + \varphi_i(x_{i+1})\}$$

$$\simeq \frac{1}{2}\{0 + 1\} - \frac{1}{2}\{1 + 0\} = 0. \tag{6.122}$$

b) *Calculation of coefficient $A_{i,i-1}$.*

$$A_{i,i-1} = \int_0^1 \varphi_i \varphi'_{i-1} \, dx = \int_{\text{Supp } \varphi_i \, \cap \, \text{Supp } \varphi'_{i-1}} \varphi_i \varphi'_{i-1} \, dx = \int_{x_{i-1}}^{x_i} \varphi_i \varphi'_{i-1} \, dx$$

$$\simeq -\frac{1}{h} \times \frac{h}{2} \{\varphi_i(x_{i-1}) + \varphi_i(x_i)\}$$

$$\simeq -\frac{1}{2} \{0 + 1\} = -\frac{1}{2} \,. \tag{6.123}$$

c) *Calculation of coefficient $A_{i,i+1}$.*

$$A_{i,i+1} = \int_0^1 \varphi_i \varphi'_{i+1} \, dx = \int_{\text{Supp } \varphi_i \, \cap \, \text{Supp } \varphi'_{i+1}} \varphi_i \varphi'_{i+1} \, dx = \int_{x_i}^{x_{i+1}} \varphi_i \varphi'_{i+1} \, dx$$

$$\simeq +\frac{1}{h} \times \frac{h}{2} \{\varphi_i(x_i) + \varphi_i(x_{i+1})\}$$

$$\simeq \frac{1}{2} \{1 + 0\} = +\frac{1}{2} \,. \tag{6.124}$$

▶ **Approximate Calculation of Coefficients B_{ijk}, $(j,k) \in \{i-1, i, i+1\}$**

According to the general expression of B_{ijk} (6.121) and given the characteristic property of each of the basis functions φ_i, $(\varphi_i(x_j) = \delta_{ij})$, only coefficient $B_{i,i,i}$ may be non-zero.

In fact, given the use of the trapezium rule to evaluate the integrals of each coefficient B_{ijk}, expressions containing terms in the form below would be systematically obtained:

$$\varphi_i(x_l) \varphi_j(x_l) \varphi_k(x_l) \,.$$

That is why all coefficients B_{ijk} will be zero by approximation *via* the trapezium rule, with the exception of $B_{i,i,i}$:

$$B_{i,i,i} = \int_{x_{i-1}}^{x_i} \varphi_i^3(x) \, dx + \int_{x_i}^{x_{i+1}} \varphi_i^3(x) \, dx \simeq \frac{h}{2} \{0 + 1\} + \frac{h}{2} \{1 + 0\} = h \,. \tag{6.125}$$

▶ **Approximate Calculation of Coefficient C_i**

The second member C_i will be evaluated according an analogous technique:

$$C_i = \int_0^1 f \varphi_i(x) \, dx = \int_{x_{i-1}}^{x_i} f \varphi_i(x) \, dx + \int_{x_i}^{x_{i+1}} f \varphi_i(x) \, dx$$

$$\simeq \frac{h}{2} \{0 + f_i\} + \frac{h}{2} \{f_i + 0\} = h f_i \,, \tag{6.126}$$

where f_i represents the value of function f at node x_i.

A.5) The previous calculations enable the writing of the corresponding nodal equation for every characteristic basis function φ_i of a node interior to the mesh:

$$\frac{1}{2}[\tilde{u}_{i+1} - \tilde{u}_{i-1}] + h\tilde{u}_i^2 = hf_i \, , \ (i = 1 \text{ to } N) \, . \tag{6.127}$$

A.6) Nodal equation (6.127) corresponds exactly to the centred finite differences scheme associated with the continuous problem **(CP)**.

In fact, this result is immediate because it consists in writing the Riccati differential equation (6.108) at point x_i, then in replacing the first derivative of u by a second-order approximation:

$$u'(x_i) + u^2(x_i) = f(x_i) \quad \text{at} \quad (i = 1, N) \, . \tag{6.128}$$

Taylor's formula is then simultaneously used progressively and regressively:

$$u(x_{i+1}) = u(x_i) + hu'(x_i) + \frac{h^2}{2}u''(x_i) + O(h^3) \, , \tag{6.129}$$

$$u(x_{i-1}) = u(x_i) - hu'(x_i) + \frac{h^2}{2}u''(x_i) + O(h^3) \, . \tag{6.130}$$

Equation (6.130) is subtracted from equation (6.129) and it becomes:

$$u'(x_i) = \frac{1}{2h}[u(x_{i+1}) - u(x_{i-1})] + O(h^2) \, . \tag{6.131}$$

Finally, by injecting the previous result in equation (6.129), it becomes:

$$\frac{1}{2h}[u(x_{i+1}) - u(x_{i-1})] + O(h^2) + u(x_i)^2 = f(x_i) \, , \ (i = 1 \text{ at } N) \, . \tag{6.132}$$

As usual, for a discretisation by finite differences, the infinitesimal term (here, $O(h^2)$) is neglected and this imbalances equation (6.132):

$$\frac{1}{2h}[u(x_{i+1}) - u(x_{i-1})] + u^2(x_i) \simeq f(x_i) \, , \ (i = 1 \text{ at } N) \, . \tag{6.133}$$

In order to restore a real equality, the sequence of unknown quantities $(u_i)_{i=1,N}$ are replaced by the numerical approximation sequence $(\tilde{u}_i)_{i=1,N}$ defined by the recurrence relationship:

$$\frac{1}{2h}[\tilde{u}_{i+1} - \tilde{u}_{i-1}] + h\tilde{u}_i^2 = hf_i \, , \ (i = 1 \text{ at } N) \, . \tag{6.134}$$

Of course, the approximation sequence $(\tilde{u}_i)_{i=1,N}$ has all reasons to produce a satisfactory approximation of the values of $(u_i)_{i=1,N}$ in so far as its construction procedure is directly motivated by that of real values $(u_i)_{i=1,N}$, (to the nearest error $O(h^2)$).

In conclusion, the finite differences scheme (6.134) corresponds exactly to nodal equation (6.127) that was obtained for any characteristic function φ_i, of a node strictly interior at the $[0, 1]$ integration interval.

▶ **Characteristic Basis Function φ_{N+1} of the Node at Abscissa x_{N+1}**

A.7) The characteristic basis function of the node at abscissa $x = 1$ is now considered; it consists of a basis function φ_{N+1}.

In this case and by a reasoning similar to the one developed in the previous question for all nodes interior at 0,1, the approximate variational formulation $\widetilde{(\text{PV})}$ leads to equation $\widetilde{(\text{PV}_{\text{Ext}})}$:

$$\widetilde{(\text{PV}_{\text{Ext}})} \begin{cases} A_{N+1,N}\tilde{u}_N + A_{N+1,N+1}\tilde{u}_{N+1} + B_{N+1,N,N}\tilde{u}_N^2 + \dots \\ (B_{N+1,N,N+1} + B_{N+1,N+1,N})\tilde{u}_N\tilde{u}_{N+1} + \dots \\ B_{N+1,N+1,N+1}\tilde{u}_{N+1}^2 = C_{N+1} \,. \end{cases} \tag{6.135}$$

Once again, the property of the supports of basis functions φ_N and φ_{N+1} would have been used.

The calculation of coefficients is then carried out as described below.

▶ **Approximate Calculation of Coefficients $A_{N+1,N}$ and $A_{N+1,N+1}$**

a) *Calculation of $A_{N+1,N}$.*

$$\begin{aligned} A_{N+1,N} &= \int_0^1 \varphi_{N+1}\varphi_N'\, dx = \int_{\text{Supp } \varphi_{N+1} \cap \text{Supp } \varphi_N'} \varphi_{N+1}\varphi_N'\, dx \\ &= \int_{x_N}^{x_{N+1}} \varphi_{N+1}\varphi_N'\, dx \\ &\simeq -\frac{1}{h} \times \frac{h}{2}\left\{\varphi_{N+1}(x_N) + \varphi_{N+1}(x_{N+1})\right\} \\ &\simeq -\frac{1}{2}\{0+1\} = -\frac{1}{2}\,. \end{aligned} \tag{6.136}$$

b) *Calculation of $A_{N+1,N+1}$.*

$$\begin{aligned} A_{N+1,N+1} &= \int_0^1 \varphi_{N+1}\varphi_{N+1}'\, dx = \int_{\text{Supp } \varphi_{N+1}} \varphi_{N+1}\varphi_{N+1}'\, dx \\ &= \int_{x_N}^{x_{N+1}} \varphi_{N+1}\varphi_{N+1}'\, dx \\ &\simeq \frac{1}{h} \times \frac{h}{2}\left\{\varphi_{N+1}(x_N) + \varphi_{N+1}(x_{N+1})\right\} \\ &\simeq \frac{1}{2}\{0+1\} = \frac{1}{2}\,. \end{aligned} \tag{6.137}$$

▶ **Approximate Calculation of Coefficients $B_{N+1,N,N+1}$, $B_{N+1,N+1,N}$ and $B_{N+1,N+1,N+1}$**

Given that the trapezium quadrature method is used and the same reasoning as for question **4** is applied then only coefficient $B_{N+1,N+1,N+1}$ may be non-zero.

The following is thus obtained:

$$B_{N+1,N+1,N+1} = \int_{x_N}^{x_{N+1}} \varphi_{N+1}^3(x)\,dx \simeq \frac{h}{2}\{1+0\} = \frac{h}{2}. \qquad (6.138)$$

▶ **Approximate Calculation of Coefficients C_{N+1}**

$$C_{N+1} = \int_0^1 f\varphi_{N+1}(x)\,dx = \int_{x_N}^{x_{N+1}} f\varphi_{N+1}(x)\,dx$$

$$\simeq \frac{h}{2}\{0 + f_{N+1}\} = hf_{N+1}, \qquad (6.139)$$

where f_{N+1} represents the value of function f at node $x_{N+1} = 1$.

A.8) The previous calculations are now grouped in order to build the corresponding nodal equation:

$$\frac{1}{h}[\tilde{u}_{N+1} - \tilde{u}_N] + h\tilde{u}_{N+1}^2 = hf_{N+1}. \qquad (6.140)$$

A.9) Nodal equation (6.140) corresponds exactly to the equation resulting from the discretisation of the Riccati differential equation (6.108) using the finite differences method.

By the way, a major difference occurs with respect to finite difference scheme (6.134) established in question **4** for the points strictly interior at the mesh.

In fact, in the present case, it is immediately observed that finite difference scheme (6.140) is no longer of the second order but only of the first order. This results in a decrease of the total order of the method since the finite differences scheme is globally of the first order.

Such a result will not be surprising if the Bramble-Hilbert lemma (cf. D. Euvrard, [4]) is kept in mind. It holds that the finite elements method is of the k-order if the approximation variational space $\tilde{V}$ contains space P_k, (all the polynomials having degrees inferior or equal to k).

In the present case, it is obvious that k is equal to one.

References

1. H. Brézis – *Analyse fonctionnelle, théorie et applications,* Masson, 1983.
2. M. Crouzeix and A. L. Mignot – *Analyse numérique des équations différentielles,* Masson, 1983.
3. Roger Dautray and Jacques-Louis Lions – *Analyse mathématique et calcul numérique pour les sciences et les techniques,* Masson, (1987).
4. G. Duvaut – *Mécanique des Milieux Continus,* Dunod, (1998).
5. D. Euvrard – *Résolution des équations aux dérivées partielles de la physique, de la mécanique et des sciences de l'ingénieur,* Masson, (1994).
6. M. Moussaoui, *Sur l'approximation des solutions du problème de Dirichlet dans un ouvert avec coins.* "Singularities and constructive methods for their treatment" P. Grisvard W. Wendland, J.R.Whiteman Editors. Lecture Notes in Math. Springer Verlag n° 1121. 1984.
7. P.A. Raviart and E. Godlewski, Numerical approximation of hyperbolic systems of conservation laws, Appl. Math. Sci., vol. 118, Springer Verlag New York, (1996).
8. P.A Raviart and J.M. Thomas – *Introduction à l'analyse numérique des équations aux dérivées partielles,* Masson, (1988).
9. L. Schwartz – *Méthodes mathématiques pour les sciences physiques,* Hermann, (1983).

Index